VOYAGE
ACROSS THE
COSMOS

GILES SPARROW

Quercus

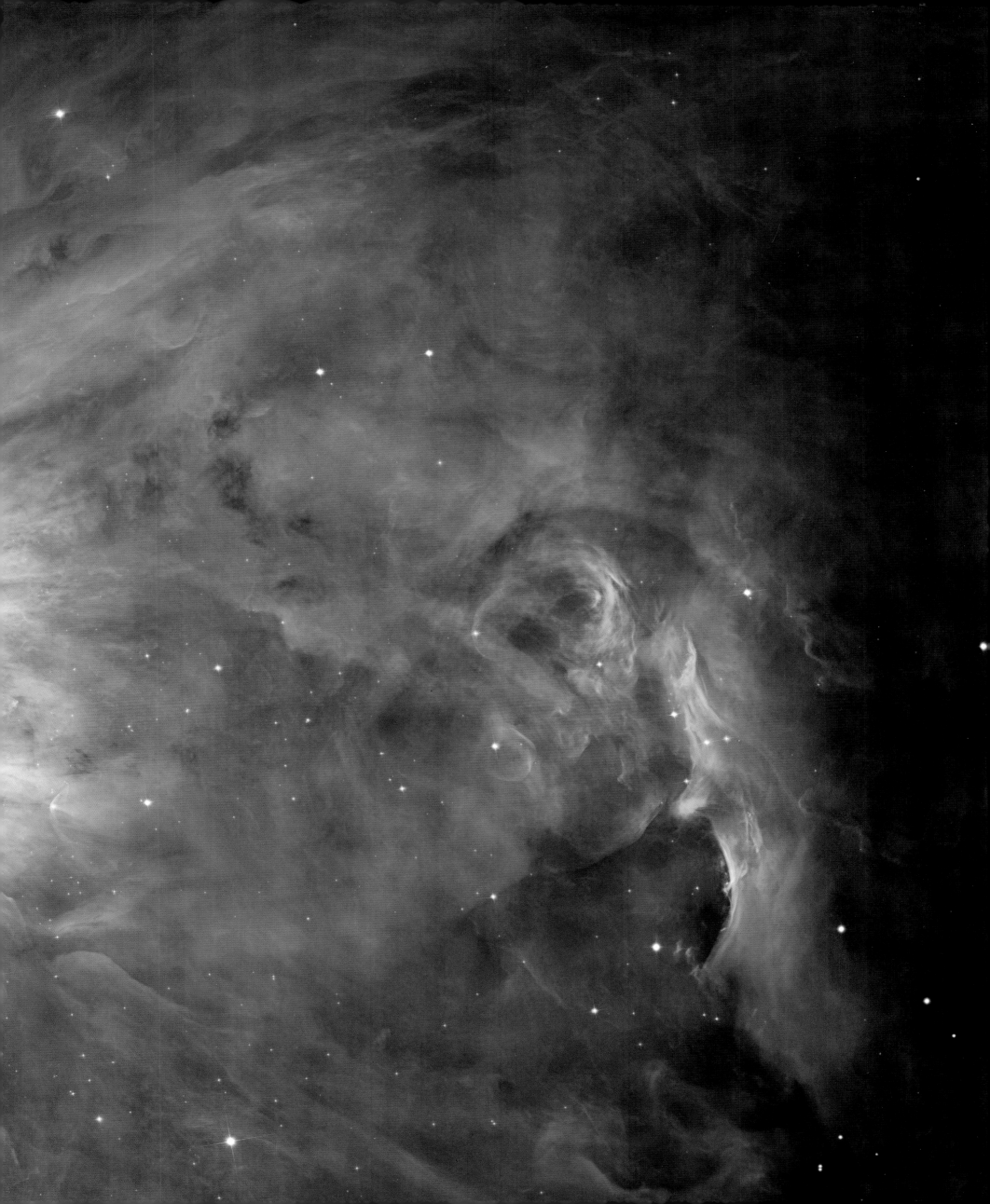

CONTENTS

ACROSS THE SOLAR SYSTEM

THROUGH THE MILKY WAY

BEYOND OUR GALAXY

DEPTHS OF THE UNIVERSE

PRE-FLIGHT INSTRUCTIONS

STRAP YOURSELF IN AND PREPARE FOR BLAST OFF – WE'RE GOING ON AN EPIC JOURNEY THROUGH THE COSMOS, ACROSS 13.7 BILLION LIGHT YEARS OF SPACE AND AS MANY YEARS OF HISTORY TO THE EDGE OF THE UNIVERSE AND THE DAWN OF CREATION. ALONG THE WAY, WE'LL VISIT THE PLANETS AND MOONS OF THE SOLAR SYSTEM, JOURNEY AMONG THE STARS OF THE MILKY WAY AND VISIT GALAXIES FAR BEYOND OUR OWN. BUT FIRST, HERE'S A GUIDE TO THE VARIOUS DATA STREAMS THAT WILL TELL YOU ABOUT THE OBJECTS WE ENCOUNTER ALONG THE WAY.

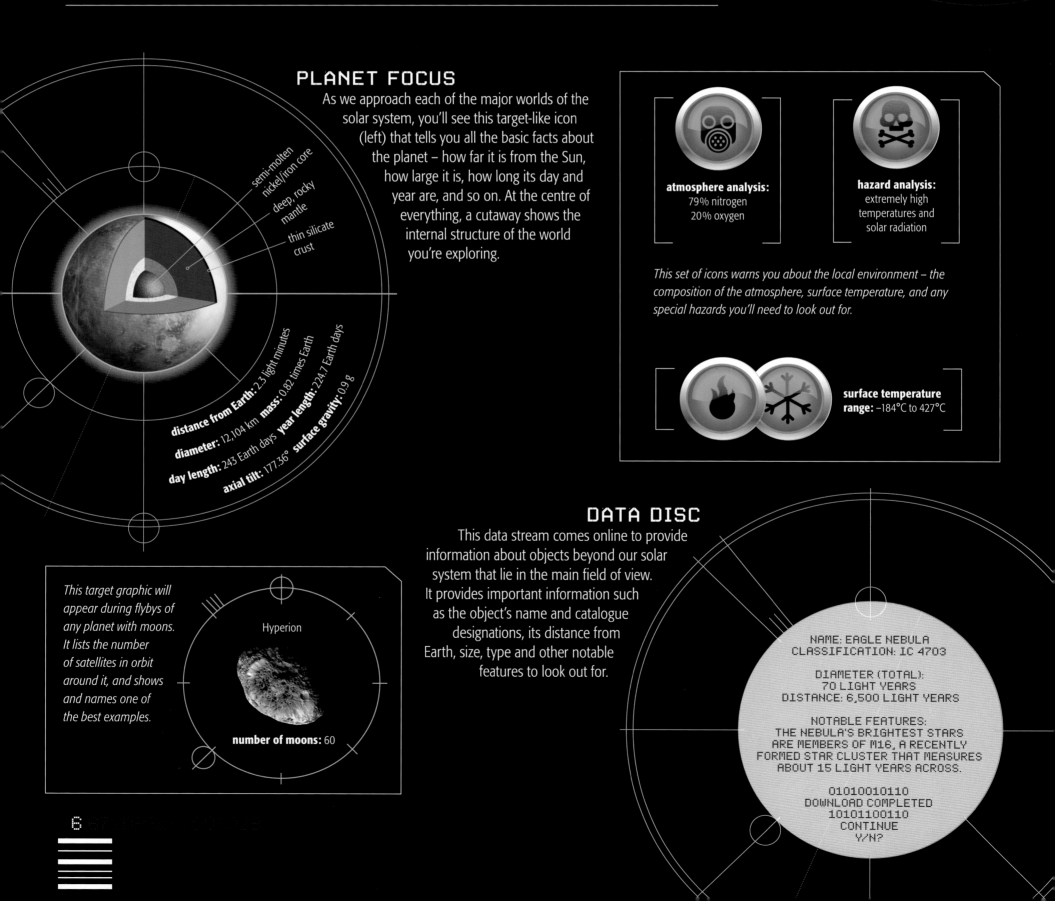

PLANET FOCUS

As we approach each of the major worlds of the solar system, you'll see this target-like icon (left) that tells you all the basic facts about the planet – how far it is from the Sun, how large it is, how long its day and year are, and so on. At the centre of everything, a cutaway shows the internal structure of the world you're exploring.

semi-molten nickel/iron core

deep, rocky mantle

thin silicate crust

distance from Earth: 2.3 light minutes
diameter: 12,104 km mass: 0.82 times Earth
day length: 243 Earth days year length: 224.7 Earth days
axial tilt: 177.36° surface gravity: 0.9 g

atmosphere analysis:
79% nitrogen
20% oxygen

hazard analysis:
extremely high
temperatures and
solar radiation

This set of icons warns you about the local environment – the composition of the atmosphere, surface temperature, and any special hazards you'll need to look out for.

surface temperature range: −184°C to 427°C

DATA DISC

This data stream comes online to provide information about objects beyond our solar system that lie in the main field of view. It provides important information such as the object's name and catalogue designations, its distance from Earth, size, type and other notable features to look out for.

This target graphic will appear during flybys of any planet with moons. It lists the number of satellites in orbit around it, and shows and names one of the best examples.

Hyperion

number of moons: 60

NAME: EAGLE NEBULA
CLASSIFICATION: IC 4703

DIAMETER (TOTAL):
70 LIGHT YEARS
DISTANCE: 6,500 LIGHT YEARS

NOTABLE FEATURES:
THE NEBULA'S BRIGHTEST STARS
ARE MEMBERS OF M16, A RECENTLY
FORMED STAR CLUSTER THAT MEASURES
ABOUT 15 LIGHT YEARS ACROSS.

01010010110
DOWNLOAD COMPLETED
10101100110
CONTINUE
Y/N?

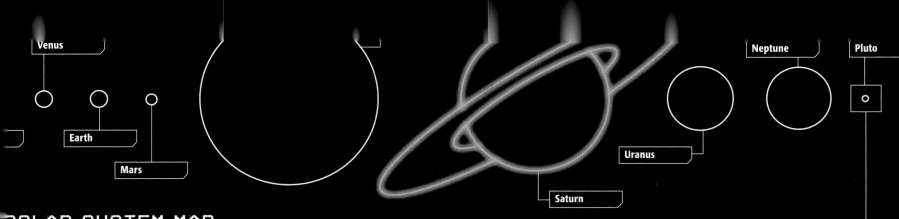

Venus

Earth

Mars

Neptune

Pluto

Uranus

Saturn

SOLAR SYSTEM MAP

This chart tracks your location among the planets of the solar system – the display appears at the start of each flyby, with your current planet highlighted. The planets are all shown approximately to scale with each other, and satellites of interest pop up on the map and are highlighted as you encounter them.

In general, the map shows just the eight major planets as defined by astronomers in 2006. The ex-planet Pluto appears only when relevant.

≪ IMAGE ENHANCE

> ## RING RAINBOW

DATA PACKET ACCEPTED. 0100101

All the giant planets have ring systems, but only Saturn's are so grand. This is partly because they contain so much material, but also because they are largely made from frozen water ice and so reflect more sunlight than rings of other material. This "rainbow" recorded by the Cassini probe is a result of sunlight being reflected back off a point directly opposite the Sun as the fast-moving probe took images of the rings through three coloured filters.

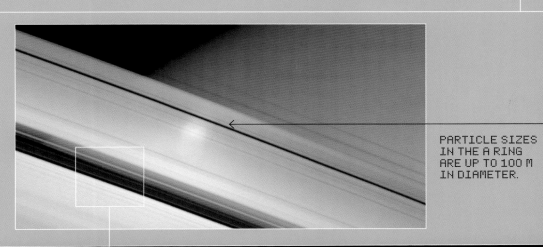

PARTICLE SIZES IN THE A RING ARE UP TO 100 M IN DIAMETER.

Various other data boxes will pop up along the journey. "Image Enhance" boxes show you details of a target object, including zoomed-in views or specially processed images that show the object in other types of radiation.

Solar system objects often display a "Surface Detail" box, with information about a particular feature or process going on within or on the surface of a planet or moon.

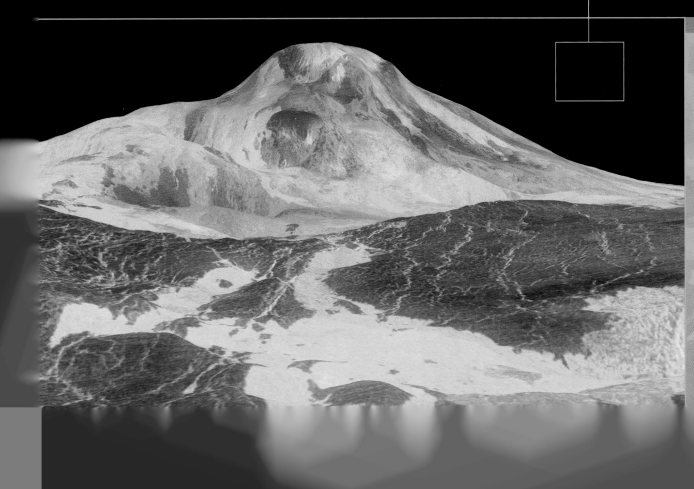

SURFACE DETAIL

RADAR MAPPING

>> Poor visibility makes it impossible to see far on the Venusian surface, while the clouds hide the terrain from orbiting satellites.

>> Radar mapping is the only way to see the real shape of the surface. Radio beams are fired downwards by an orbiting probe, and the time their echoes take to return shows how far away the surface is and, therefore, how high or low-lying.

>> This three-dimensional image of Maat Mons (left), a volcano 8 km high, is a computer reconstruction of radar data collected by NASA's Magellan spacecraft.

ACROSS THE SOLAR SYSTEM

EMPIRE OF THE SUN

Our small corner of space is ruled over by the influence of a single massive object – our local star. The Sun pumps out enormous amounts of energy that we rely on for heat and light, sends streams of dangerous high-speed particles out across nearby space, and through its powerful gravity also holds the entire solar system in orbit around it.

The solar system itself consists of eight major planets, dozens of satellites that orbit around them, and countless smaller worlds, ranging in size from substantial "dwarf planets" to mere fragments of debris. Each object (apart from the satellites) follows its own path around the Sun – an orbit that is usually an ellipse rather than a perfect circle. At its closest to the Sun (perihelion), the object moves faster along its orbit than when it is furthest away (at aphelion).

The orbits of the major planets are all more or less circular (the most elliptical are those of Mercury and Mars), and they all lie in roughly the same plane, giving the solar system the shape of a flattened disc. Smaller worlds such as asteroids and comets often follow orbits that are more elliptical, and tilted compared to the rest of the solar system.

The planets divide into two groups of four – rocky planets closer to the Sun, and giants further out. In order from the Sun, the rocky planets are Mercury, Venus, Earth and Mars. Earth is the largest of these, and the only one with its own substantial satellite, the Moon (though Mars is orbited by a pair of small, captured asteroids). Each of the rocky planets has a dense interior with a metal core at the centre. Heat from within them has powered geological forces at different rates throughout their history, and their surfaces have also been shaped by the influence of atmospheres and surface water (where they exist).

The giant planets are far larger but less substantial worlds, composed largely of gas, liquid and slushy ice. In order from the Sun, they are Jupiter (the largest planet in the solar system), Saturn, Uranus and Neptune. The giants have enormous atmospheres made of light gases that compress into liquid or ice beneath the surface, wrapped around a core the size of a rocky planet. All of the giant planets are surrounded by rings – streams of small particles trapped in orbit – and each has a large family of moons, some of which are almost the size of rocky planets.

The debris of the solar system consists of rocky fragments called asteroids closer to the Sun, and icy comets and "ice dwarfs" further out (though some comets fall into elliptical orbits that bring them close to the Sun). The asteroids are mostly confined in a doughnut-shaped ring, the Asteroid Belt, between Mars and Jupiter. Ice dwarfs occupy a similar but looser ring, the Kuiper Belt, beyond the orbit of Neptune. Some comets also lurk in the Kuiper Belt, but most stay much further out, in a huge spherical shell called the Oort Cloud, at the very limits of the Sun's gravitational reach.

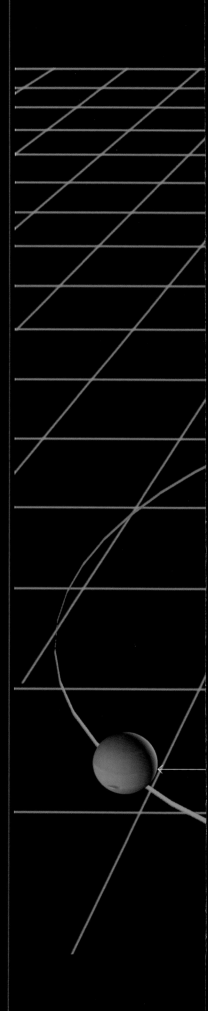

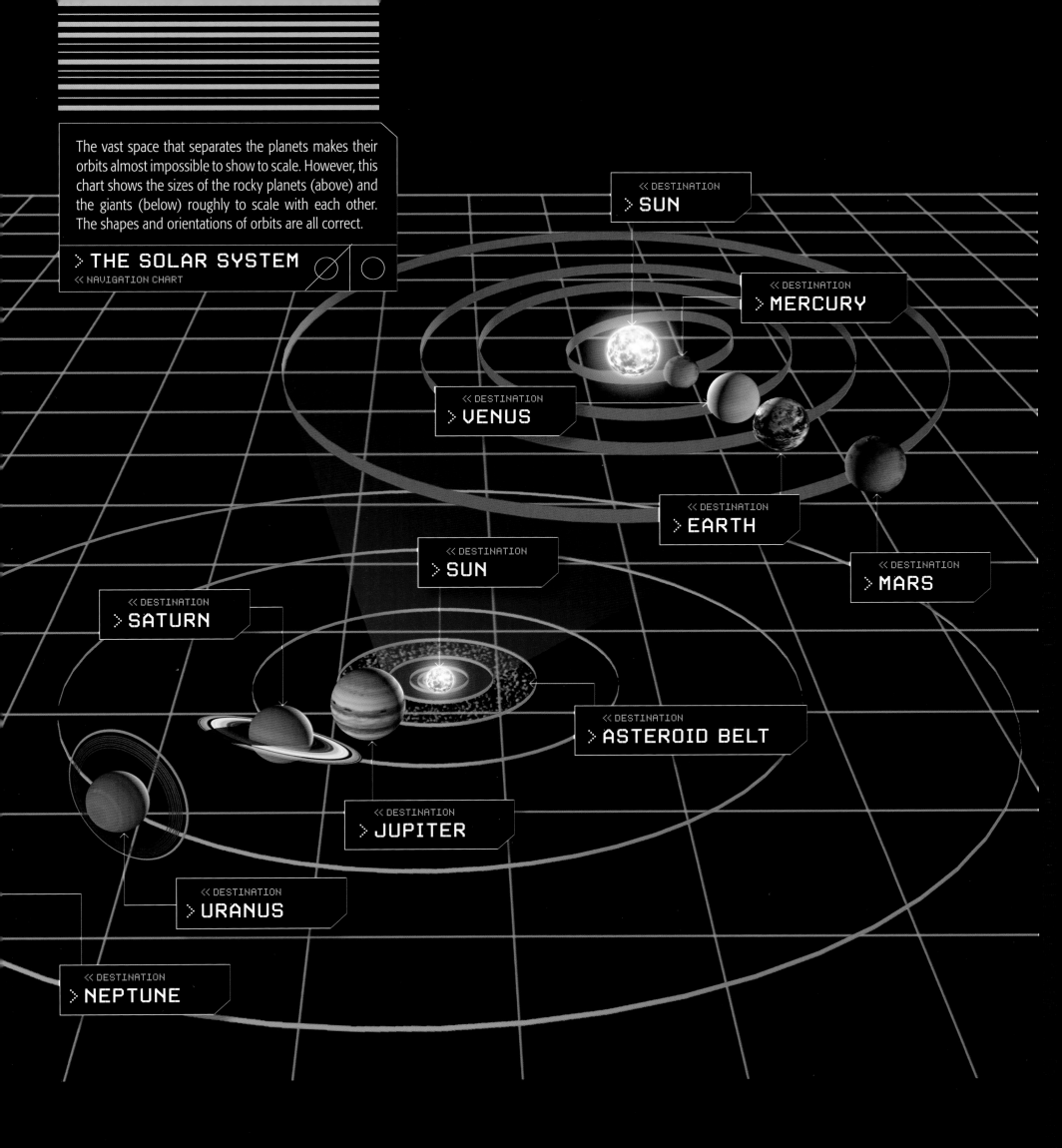

The vast space that separates the planets makes their orbits almost impossible to show to scale. However, this chart shows the sizes of the rocky planets (above) and the giants (below) roughly to scale with each other. The shapes and orientations of orbits are all correct.

> THE SOLAR SYSTEM
<< NAVIGATION CHART

<< DESTINATION
> SUN

<< DESTINATION
> MERCURY

<< DESTINATION
> VENUS

<< DESTINATION
> EARTH

<< DESTINATION
> MARS

<< DESTINATION
> SUN

<< DESTINATION
> SATURN

<< DESTINATION
> ASTEROID BELT

<< DESTINATION
> JUPITER

<< DESTINATION
> URANUS

<< DESTINATION
> NEPTUNE

EARTH

WITH ITS BLUE OCEANS AND WHITE CLOUDS, PLANET EARTH LOOKS LIKE A GIANT BLUE-AND-WHITE MARBLE HANGING IN THE DARKNESS OF SPACE. BUT OUR WORLD IS MAINLY MADE OF ROCK AND IS THE LARGEST OF THE FOUR TERRESTRIAL PLANETS IN THE SOLAR SYSTEM.

solid inner nickel/iron core
molten outer nickel/iron core
silicate mantle
crust

diameter: 12,756 km **mass:** 5.97 trillion trillion kg

day length: 23 hours 56 minutes **year length:** 365.25 days

axial tilt: 23.45° **surface gravity:** 1 g

surface temperature range: −15°C to 37°C

hazard analysis: hospitable planet, extreme temperatures at poles

atmosphere analysis:
79% nitrogen
20% oxygen

The Moon

number of moons: 1

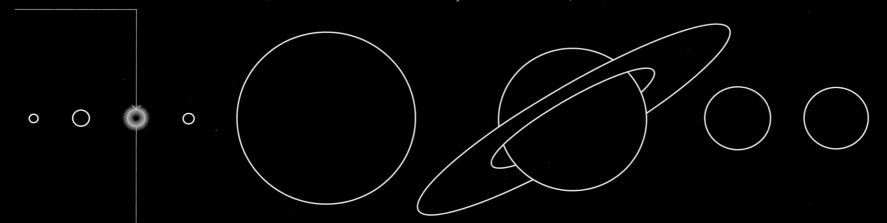

<< IMAGE ENHANCE

EARTH'S ATMOSPHERE

The atmosphere covers Earth with a blanket of gases that extends for more than 100 km into space but is only substantial for the first 10 km. Gases in the atmosphere are vital to life on Earth and protect us from deadly space radiation and high-energy particles. They also help to distribute heat around the world.

Clouds drawn into areas of low pressure form swirls called cyclones and anticyclones.

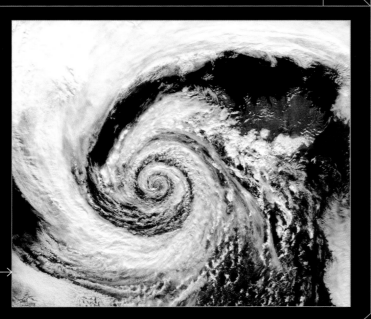

Earth's poles receive less heat and light from the Sun than its equator, so they are almost permanently frozen. Ice around the North Pole floats on the Arctic Ocean as pack ice (above). Around the South Pole, it builds up on and around the continent of Antarctica.

> **POLAR ICE**
<< IMAGE ENHANCE

SURFACE DETAIL
OCEAN PLANET

>> More than 70 per cent of Earth's surface is covered in oceans, and more than half of this area is at least 3 km deep. Having liquid surface water makes our planet unique in the solar system. It affects the planet's climate, weather and even the movement of its rocks. This image shows part of the Grand Bahama Canyon, a deep oceanic trench.

>> Circulation of the water in the oceans mirrors that of air in the atmosphere, carrying warm water away from the equator and towards the poles in enormous "gyres" that turn anticlockwise in the northern hemisphere and clockwise in the southern hemisphere.

>> Deep undersea troughs and volcanic ridges mark the edges of the ocean plates that make up the sea floor. In some places, the troughs are up to 11 km deep.

NAME: HIMALAYAS

TYPE: MOUNTAIN RANGE

DIMENSIONS: 3,000 KM LONG
UP TO 8,800 M HIGH

NOTES:
ZONE OF TECTONIC COLLISION BETWEEN
TWO CONTINENTAL PLATES, STILL BEING
PUSHED UPWARDS BY 5 MM PER YEAR.

01010010110
DOWNLOAD COMPLETED
10101100110
CONTINUE
Y/N?

ꟶ TECTONIC PLATES

Tectonic plates come in two varieties. Thick, old continental plates contain the continents and often a fringe of thinner, younger material. Thin, young oceanic plates lie mostly beneath the seas. New material is formed by volcanoes at places where the plates are separating. Where oceanic plates are colliding with continental ones, the oceanic material is forced back down into the mantle.

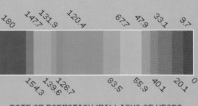

DATE OF FORMATION (MILLIONS OF YEARS
BEFORE PRESENT)

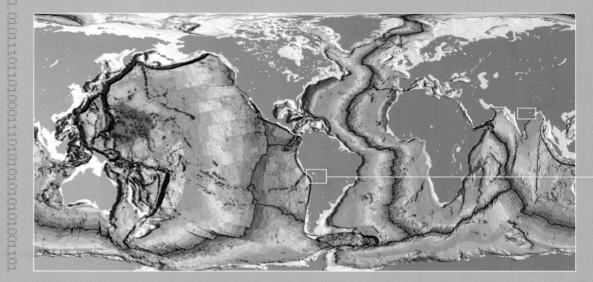

DATA PACKET ACCEPTED..01011011010001110101001010101001101.

872DARWY7001044

EARTH TECTONICS

INCOMING DATA... ACCEPTED >

EARTH'S OUTER CRUST IS SPLIT INTO DOZENS OF LARGE AND SMALL CHUNKS CALLED TECTONIC PLATES. THESE ARE PUSHED AND PULLED BACK AND FORTH ON A FLUID LAYER OF THE UPPER MANTLE CALLED THE AESTHENOSPHERE. INTERACTION BETWEEN THE PLATES FUELS MUCH OF OUR PLANET'S GEOLOGICAL ACTIVITY.

« IMAGE ENHANCE

> VOLCANOES

EXPAND SEARCH »

101010011000

DATA PACKET ACCEPTED..0101011101011010001

Volcanoes mark spots where hot material in Earth's mantle is forcing its way to the surface. Beneath the ground, this material is known as magma; above the ground, it forms lava, which solidifies into new rock. Volcanoes are often found in huge belts on the edge of continents, such as in the Andes Mountains, where an ocean plate is melting as it is pushed beneath the South American continent, releasing heat in the process. In some places, such as Hawaii, "hot spots" in the mantle simply burn their way up through the thin oceanic crust.

NAME: PAMPAS LUXSAR
LOCATION: ANDES MOUNTAINS
SCALE: 60 KM BY 60 KM
MISSION: TERRA SATELLITE

01010010110
DOWNLOAD COMPLETED
10101100110

THE MOON

INCOMING DATA... ACCEPTED >

EARTH'S HUGE SATELLITE IS A QUARTER OF THE SIZE OF ITS PARENT PLANET. IT FORMED FROM MATERIAL THROWN OFF IN AN INTERPLANETARY COLLISION WHEN EARTH WAS YOUNG. ITS SURFACE IS DOMINATED BY BRIGHT, CRATERED HIGHLANDS AND DARK, SMOOTH "SEAS" – THE REMAINS OF ANCIENT LAVA ERUPTIONS.

small solid nickel/iron core

deep silicate mantle

crust

distance from Earth: 384,400 km

diameter: 3,476 km   **mass:** 0.012 times Earth   **orbital period:** 27.32 Earth days

rotation period: 27.32 Earth days   **axial tilt:** 1.54°   **surface gravity:** 0.17 g

surface temperature range: –233°C to 123°C

hazard analysis: solar radiation, extreme temperatures

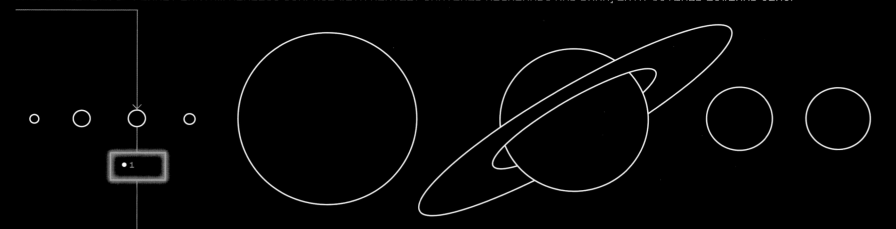

• 1

« IMAGE ENHANCE

› FAR SIDE

The Moon spins once in each orbit, which means it keeps the same face permanently turned towards us. The Moon's far side is quite different from the near side. It has cratered highlands and deep basins formed by huge impacts, but it never experienced the widespread lava eruptions that created the seas on the near side. As a result, the far side has just a few small, dark seas.

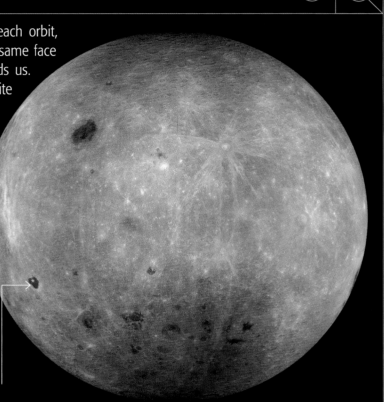

Tsiolkovskii is a far-side crater flooded with dark lava.

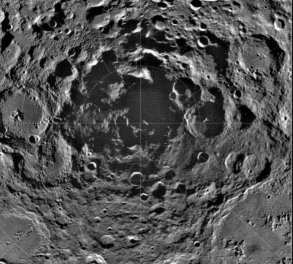

A huge impact crater, the largest in the solar system, lies hidden at the Moon's south pole. The South Pole-Aitken Basin is the size of western Europe and is mostly on the Moon's far side. It is filled with younger craters, some of which are so deep that they never see sunlight.

› SOUTH POLE

« IMAGE ENHANCE

SURFACE DETAIL
CRATERS

›› The craters that cover the Moon were mostly formed by meteorites hitting the surface. Copernicus (left) is a medium-sized crater, 91 km across and 900 million years old.

›› The largest lunar craters formed the basins in which the seas are now found about 3.9 billion years ago, during an event called the Late Heavy Bombardment. This was when large objects orbiting between the planets were "soaked up".

›› Craters are still forming today. Collisions have made new craters in the smooth lunar seas.

›› The youngest craters are surrounded by rays of bright material, stretching for hundreds of kilometres across the Moon, where debris was scattered on impact.

> APOLLO LANDER

DATA PACKET ACCEPTED. 010010011.01

The spider-like Apollo Lunar Module (LM) was designed to fly only in space and land on the lunar surface. For launch, it was shrouded in a protective cover on top of an enormous Saturn V rocket. Once in orbit around the Moon, the LM separated from the rest of the spacecraft and flew to the surface, slowing its descent using a built-in rocket engine. For the return trip, the lower half of the LM acted as a launch pad for the upper section to blast off using its own small rocket and return to orbit.

APOLLO MISSIONS

INCOMING DATA... ACCEPTED >

BETWEEN 1969 AND 1972, TWELVE AMERICAN ASTRONAUTS WALKED ON THE LUNAR SURFACE. APOLLO 11 WAS THE FIRST MISSION TO LAND ON THE MOON. FIVE OTHER APOLLOS, 12 TO 17 (BUT NOT 13), ALSO LANDED AND CARRIED OUT SCIENCE EXPERIMENTS THAT REVEALED MANY OF THE MOON'S SECRETS.

NAME: TAURUS-LITTROW VALLEY

TYPE: HIGHLAND/SEA BOUNDARY

MISSION: APOLLO 17

DATE: DECEMBER 1972

NOTES: ASTRONAUT SCHMIDT STANDS BY THE LUNAR ROVING VEHICLE (LRV) AT THE EDGE OF A CRATER. THE LRV WAS CARRIED ABOARD APOLLO MISSIONS 15 TO 17.

01010010110
DOWNLOAD COMPLETED
10101100110
CONTINUE
Y/N?

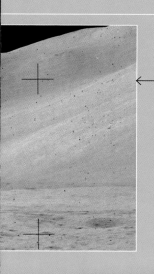

NAME: LUNAR MODULE
FALCON
LOCATION: MONS
HADLEY, LUNAR
APPENINES
DATE: 31 JULY 1971
MISSION: APOLLO 15

01010010110
DOWNLOAD COMPLETED
10101100110
CONTINUE
Y/N?
10101100101
10101100101

DATA PACKET ACCEPTED. 010101101010001110110001110101010101010101001

At 2.56 GMT on 21 July 1969, Neil Armstrong became the first human to set foot on the Moon. The Apollo 11 LM landed in the Sea of Tranquility carrying Armstrong, the mission commander, and his LM pilot, Buzz Aldrin. The grey lunar soil, formed by countless microscopic impacts throughout the Moon's history, was powdery but surprisingly stable. Footprints left behind by the astronauts should remain unchanged for many millions of years.

OBJECT TYPE:
ASTRONAUT FOOTPRINT
LOCATION: SEA OF
TRANQUILITY
DATE: 21 JULY 1969
MISSION: APOLLO 11

01010010110
DOWNLOAD COMPLETED
10101100110
CONTINUE
Y/N?
10101100101
10101100101

> FOOTSTEPS
<< IMAGE ENHANCE

VENUS

VENUS IS THE CLOSEST PLANET TO EARTH AND IS ITS NEAR-TWIN IN TERMS OF SIZE, BUT ITS APPARENT SIMILARITIES HIDE DEADLY DIFFERENCES. VENUS IS A PARCHED VOLCANIC WORLD HIDDEN BENEATH A DENSE, TOXIC ATMOSPHERE THAT TRAPS HEAT, CREATING A FIERCE GREENHOUSE EFFECT.

semi-molten nickel/iron core

deep, rocky mantle

thin silicate crust

distance from Earth: 2.3 light minutes **mass:** 0.82 times Earth

diameter: 12,104 km **year length:** 224.7 Earth days

day length: 243 Earth days **surface gravity:** 0.9 g

axial tilt: 177.36°

mean surface temperature: 464°C

atmosphere analysis:
97% carbon dioxide
3% nitrogen

hazard analysis:
extremely high
temperatures and
pressures, corrosive
vapours

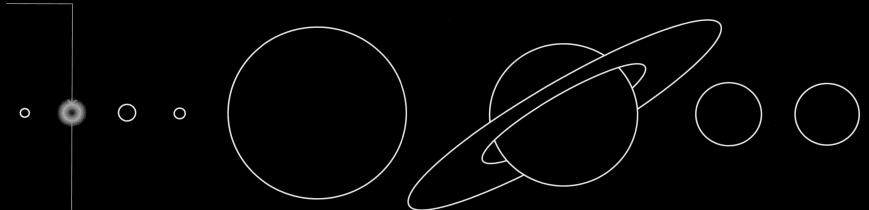

<< IMAGE ENHANCE
> HOSTILE SURFACE

The atmosphere of Venus is so dense, hot and corrosive that early attempts at landing on the planet by robot space probes failed before reaching the surface. Anything attempting a landing is simultaneously crushed by air pressure a hundred times greater than Earth's, scorched by temperatures hot enough to melt lead and burned by searing sulphuric acid vapour. The Russian probe Venera 9 was the first mission to successfully send back pictures from the surface in 1975, and Venera 13 returned this colour image in 1982 (below). Most of the surface around the landing site was composed of flat slabs of volcanic rock on dark soil. Despite heavy shielding, the probe only operated on the surface for two hours and seven minutes.

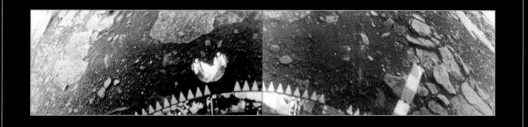

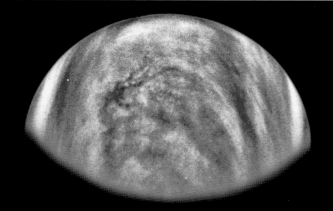

Dazzling high-level clouds bounce back 60 per cent of the light from the Sun and make Venus the most reflective planet in the solar system. They also trap heat close to the planet's surface, creating a greenhouse effect. Ultraviolet filters reveal a huge chevron-shaped pattern of clouds that circles the planet every four days.

> CLOUDS
<< RADIATION ANALYSIS

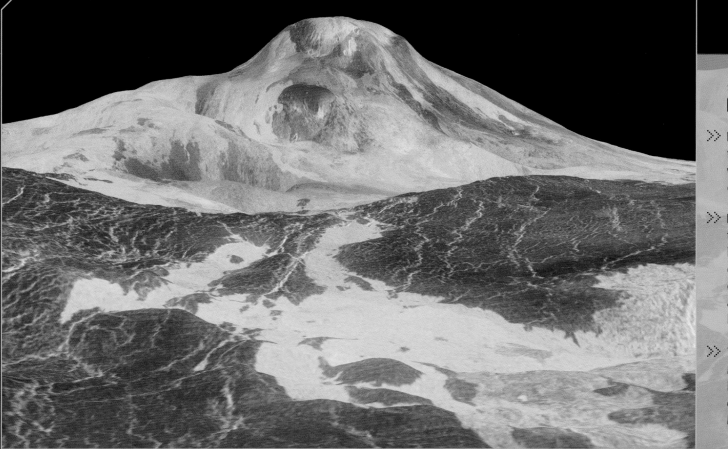

SURFACE DETAIL
RADAR MAPPING

>> Poor visibility makes it impossible to see far on the Venusian surface, while the clouds hide the terrain from orbiting satellites.

>> Radar mapping is the only way to see the real shape of the surface. Radio beams are fired downwards by an orbiting probe, and the time their echoes take to return shows how far away the surface is and, therefore, how high or low-lying.

>> This three-dimensional image of Maat Mons (left), a volcano 8 km high, is a computer reconstruction of radar data collected by NASA's Magellan spacecraft.

NAME: EISTLA REGIO

TYPE: VOLCANIC PLAIN

DIMENSIONS: 8,025 KM DIAMETER

AGE: LESS THAN 500 MILLION YEARS

NOTES: LOW-LYING PLAIN FORMED BY
SOLIDIFIED LAVA FLOWS, ORIGINATING
FROM VOLCANOES SUCH AS GULA MONS
(LEFT) AND SIF MONS (RIGHT).

DOWNLOAD COMPLETED

CONTINUE
Y/N?

672DARWY7001...KD-PPE>>66371020

VENUSIAN VOLCANOES

INCOMING DATA... ACCEPTED >
A VOLCANIC LANDSCAPE COVERS ABOUT
85 PER CENT OF VENUS'S SURFACE. THERE
ARE MANY DIFFERENT TYPES OF VOLCANO
ON VENUS, AND THEY ARE ALL FAIRLY
RECENT – MOST OF THE LAND WAS
RE-COVERED BY VOLCANIC ERUPTIONS
ABOUT 500 MILLION YEARS AGO.
ASTRONOMERS ARE NOT SURE WHETHER THE
VOLCANOES ARE STILL ERUPTING TODAY.

<< IMAGE ENHANCE
> IDEM-KUVA CORONA

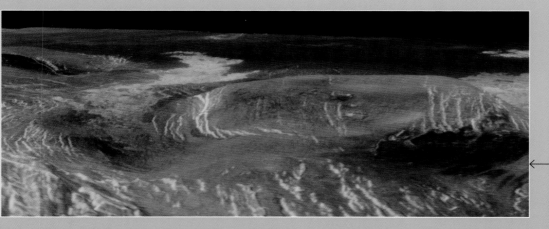

DATA PACKET ACCEPTED..01.001.01.11

Coronae are huge ring-shaped cracks in the surface, up to 100 km across. They seem to be related to novae (star-shaped cracks and ridges) and arachnoids (web-like patterns that combine the two). Coronae probably formed when molten rock beneath the crust pushed upwards until the surface cracked apart. As lava erupted through the cracks, the centres of the coronae collapsed and sank.

NAME: IDEM-KUVA CORONA
LOCATION: NORTHERN EISTLA REGIO
DIAMETER: 97 KM
MISSION: MAGELLAN

<< EXPAND SEARCH

<< EXPAND SEARCH

Pancake domes are disc-shaped pillows of solidified lava, most about 750 m high and 25 km across. They probably formed when thick lava erupted through a crack in the surface. Instead of running away from this vent, the lava piled up on top of it, solidifying and sealing the crack. Pancake domes often form in clusters, probably showing where lava was forced to find several different ways to the surface.

NAME:
PANCAKE DOMES
LOCATION:
EASTERN ALPHA
REGIO
DIAMETER:
25 KM EACH
MISSION:
MAGELLAN

01010010110
DOWNLOAD
COMPLETED

DATA PACKET ACCEPTED. 01.0001.011101000111010101.01.01000110

> ALPHA REGIO PANCAKE DOMES

<< IMAGE ENHANCE

MERCURY

INCOMING DATA... ACCEPTED >

THE INNERMOST PLANET, MERCURY, IS A SCORCHING,
AIRLESS WORLD THAT LOOKS A LITTLE LIKE THE MOON.
APART FROM THEIR CRATERS, HOWEVER, THE TWO HAVE
LITTLE IN COMMON. MERCURY'S INACCESSIBILITY
MEANS THAT IT STILL HIDES MANY SECRETS.

largely solid nickel/iron core
thin silicate mantle
crust

distance from Earth: 4.33 light minutes
mass: 0.05 times Earth
year length: 88 Earth days
diameter: 4,875 km
day length: 58.6 Earth days
axial tilt: 2°
surface gravity: 0.38 g

**surface temperature
range:** −184°C to 427°C

atmosphere analysis:
negligible – particles
of sodium, potassium
and oxygen

hazard analysis:
extremely high
temperatures and
solar radiation

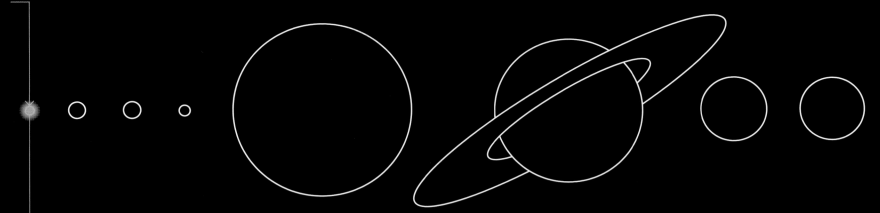

≫ CRUSTAL CRACKS

Scarps, known as "rupes", split Mercury's surface into a jigsaw, cutting across earlier features and often separating them with steep cliffs up to 2 km high. Astronomers think that Mercury swelled up early in its history, splitting its crust. As it shrank back to less than its original size, the chunks jammed awkwardly back into place.

NAME: DISCOVERY RUPES
LOCATION: MERCURY'S
SOUTHERN HEMISPHERE
DIMENSIONS: 500 KM LONG,
2 KM HIGH
MISSION: MARINER 10

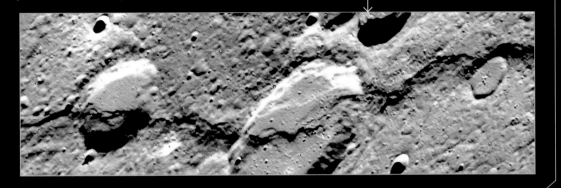

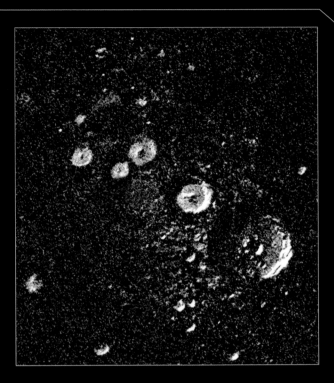

Most of Mercury is heated by the Sun to searing temperatures, but its poles get hardly any sunlight. There, radio telescopes have detected icy material in deep, dark craters. The ice may have been deposited when comets crashed into the planet.

≫ POLAR ICE

SURFACE DETAIL
CALORIS BASIN

≫ Mercury's largest feature is the Caloris Basin, an impact basin 1,350 km across. It formed when a 100-km asteroid slammed into the planet about 4 billion years ago.

≫ The basin is ringed by three mountain ranges – the remains of the crater rim. The "ejecta" material sprayed out across the surface and formed lines of hills stretching up to 1,000 km beyond the rim.

≫ Lava filled the crater, fracturing as it set to form complex patterns. Most intriguing of all is the central feature known as "the spider" (left).

≫ Shockwaves from the impact rippled round the planet – where they met again on the opposite side, they churned up the landscape to produce a chaotic region called "weird terrain".

SUN

OUR NEAREST STAR IS AN ENORMOUS BALL OF GAS
ABOUT 1.4 MILLION KM ACROSS. IT IS POWERED BY
NUCLEAR REACTIONS AT ITS CORE AND ITS EFFECT
IS FELT ACROSS THE SOLAR SYSTEM.

core

radiative zone

convective zone

photosphere

distance from Earth: 8 light minutes **mass:** 333,000 times Earth

diameter: 1.4 million km **rotation period:** 25–34 days

mean core temperature: 15 million°C

mean surface temperature: 5,500°C

atmosphere analysis:
corona of superheated
electrically charged ions

hazard analysis:
extreme temperatures
and radiation

**WARNING:
DO NOT LOOK
AT THE SUN!**

<< IMAGE ENHANCE

> SOLAR FLARES

The Sun's visible surface is the photosphere – a layer of gas at a temperature of 5,500°C. Beyond this lies the corona, an outer atmosphere extending for millions of kilometres. The corona is extremely hot but sparse and almost invisible, except when huge eruptions known as flares inject more material from below. Particles streaming out of the corona form a solar wind that stretches across the solar system.

The largest solar eruptions, known as Coronal Mass Ejections, release as much as 100 billion tonnes of matter.

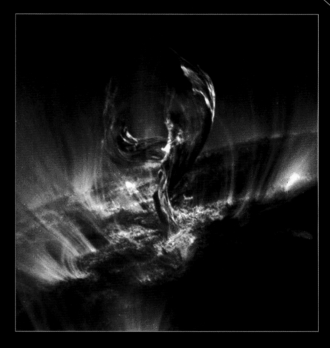

Loops in the Sun's magnetic field carry arcs of cool gas above the photosphere. When silhouetted against the hotter surface, they appear as dark filaments, but when seen above the Sun's surface, they form reddish prominences.

> PROMINENCES

<< IMAGE ENHANCE

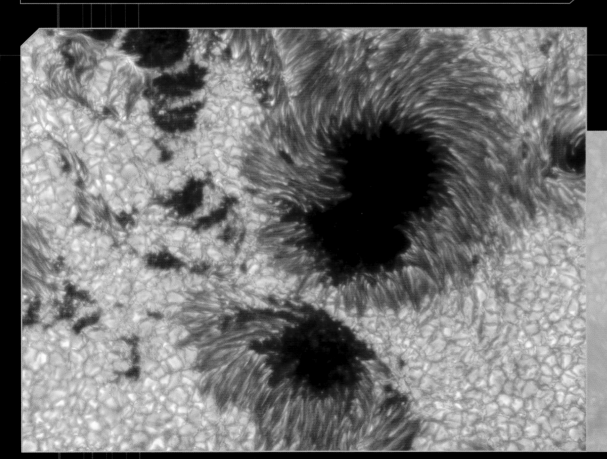

SURFACE DETAIL
SUNSPOTS

>> The bright disc of the photosphere is frequently marred by darker regions known as sunspots. They can be seen in photographs or when the Sun is safely projected onto a screen. NEVER look at the Sun directly as it could permanently damage your eyesight.

>> Sunspots mark gaps in the photosphere, where the gas is less dense thanks to an emerging loop of magnetic field. They are cooler than the rest of the photosphere, with temperatures of about 3,500°C, and so appear dark.

>> Sunspots, flares and other solar activity vary in a cycle of about 11 years, driven by changes to the Sun's internal magnetic field.

COMET FLYBY

INCOMING DATA... ACCEPTED >

CLOSE TO THE SUN, IT'S EASY TO INTERCEPT COMETS ON THEIR BRIEF VISITS. THESE IRREGULAR ICY WORLDS SPEND MOST OF THEIR TIME FAR OUT IN THE DEPTHS OF SPACE, BUT THEY FIRE INTO LIFE WHEN THEY ARE HEATED BY THE SUN.

PERIHELION: 50 LIGHT SECONDS
ORBITAL PERIOD: 37,000 EARTH YEARS
DIAMETER: UNKNOWN
01010010110
DOWNLOAD COMPLETED
10101100110
CONTINUE
Y/N?

The Near-Earth Asteroid Tracking project, NEAT, discovered this long-period comet, which comes close to the Sun just once every 37,000 years. During its last flyby, a huge gaseous coma formed around it. Pressure from the solar wind created an extended tail that pointed away from the Sun.

> **COMET NEAT**
≪ DATA FILE

2DARWY7001029110GRT5B...KD7767.2DARWY3910-PFE>>76>>

PERIHELION: 4.9 LIGHT MINUTES
ORBITAL PERIOD: 76 EARTH YEARS
LENGTH: 15.3 KM

DOWNLOAD COMPLETED

CONTINUE
Y/N?

The most famous comet of all is Halley, which orbits once every 76 years. In 1986, the Giotto probe flew past the comet's solid nucleus. Its images revealed a peanut-shaped body with a dark crust. Jets of vapour were emerging through cracks from the icy interior beneath.

> COMET HALLEY
<< DATA FILE

<< DATA FILE

> COMET TEMPEL, COMET BORRELY AND COMET WILD 2

Tempel, a short-period comet targeted by the Deep Impact space probe in 2005, orbits the Sun every 5.5 years. It once followed a much longer orbit, but it was disrupted by a close encounter with Jupiter in 1881.

Each passage around the Sun depletes the supply of ice inside a comet until eventually its activity fades. Comet Borrely's 6.9-year orbit has left it almost exhausted, with a blackened crust that makes it the darkest object in the solar system.

Although Wild 2's short orbit of 6.4 years keeps it closer to the Sun than Jupiter, its bright surface and plentiful supplies of ice suggest that it has only recently been pulled into this orbit, probably by Jupiter's own gravity.

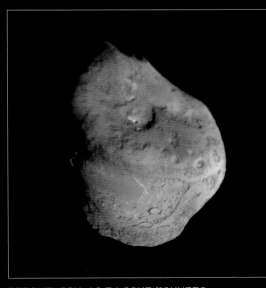

PERIHELION: 12.5 LIGHT MINUTES
ORBITAL PERIOD: 5.5 EARTH YEARS
DIAMETER: 6 KM

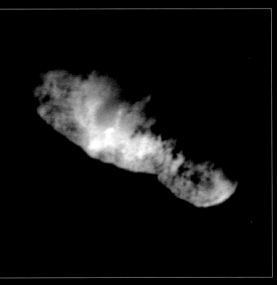

PERIHELION: 11.3 LIGHT MINUTES
ORBITAL PERIOD: 6..86 EARTH YEARS
LENGTH: 8 KM

PERIHELION: 13.1 LIGHT MINUTES
ORBITAL PERIOD: 6.39 EARTH YEARS
DIAMETER: 5.5 KM

solidified
iron core

silicate mantle

rocky crust

MARS

INCOMING DATA... ACCEPTED >

MARS IS LIKE A SMALLER, COLDER AND DRIER VERSION
OF EARTH AND IN THE PAST, IT WAS MUCH MORE
LIKE OUR OWN PLANET. TODAY, MARS IS A WORLD OF
EXTREMES, WITH TOWERING VOLCANOES, DEEP CANYONS
AND FROZEN ICE CAPS.

distance from Earth: 3.1 light minutes
diameter: 6,780 km **mass:** 0.11 times Earth
day length: 24.6 Earth hours **year length:** 687 Earth days
axial tilt: 25.19° **surface gravity:** 0.38 g

**surface temperature
range:** −125°C to 25°C

hazard analysis:
ultraviolet radiation

atmosphere analysis:
95.7% carbon dioxide
2.7% nitrogen
1.6% argon

Phobos

number of moons: 2

PLANET LOCATOR > MARS >

OUTERMOST TERRESTRIAL PLANET... THIN ATMOSPHERE AND SIGNS OF SURFACE WATER ACTIVITY AND ACTIVE VOLCANOES IN THE PAST... POLAR ICE CAPS.

<< IMAGE ENHANCE

> DUST STORMS

The Martian air is very thin, with just 0.6 per cent of Earth's atmospheric pressure, but it is dense enough to have weather. High winds can rip around the planet and stir up the fine red sands that cover much of the surface, creating enormous regional and even global dust storms. Such storms can hide the Martian surface from view for months.

Mars before (left) and during (right) a major dust storm in 2001.

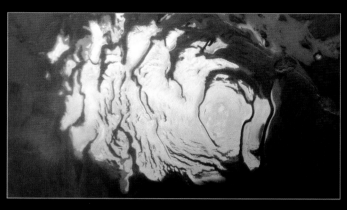

Mars has a tilted axis similar to that of Earth and so has a similar cycle of seasons throughout its year. Ice caps at the poles grow in winter and shrink in summer. Water ice and frozen carbon dioxide ("dry ice") are laid down in rippling swirls that show the direction of the prevailing winds.

> POLAR ICE CAPS

<< IMAGE ENHANCE

SURFACE DETAIL
WATER EROSION

>> Mars is covered in evidence that water once flowed across its surface. This evidence includes the scars left by sudden, catastrophic floods and the beds of sedate, long-lived rivers.

>> More than 3 billion years ago, the low-lying northern plains were covered by shallow oceans. Robot explorers have found rocks and minerals that formed under this ancient sea.

>> The gulleys around the edge of some valleys, such as Gorgonum Chaos (right), formed very recently, perhaps when water escaped from just below the surface.

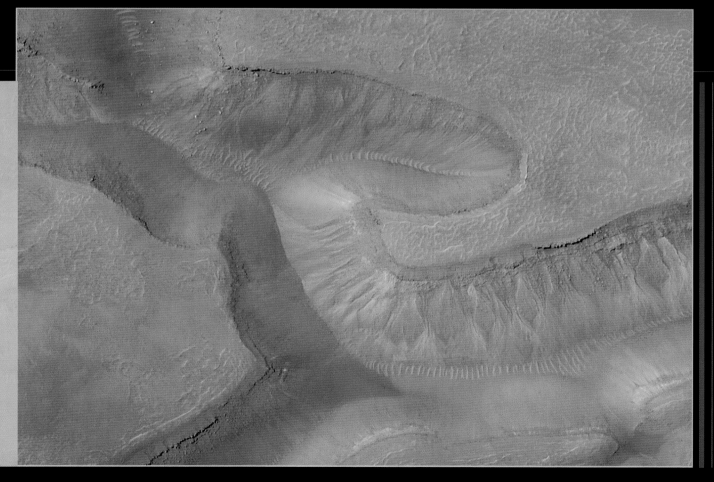

NAME: OLYMPUS MONS

HEIGHT: 27 KM

WIDTH: 550 KM

NOTABLE FEATURES:
OLYMPUS MONS IS HUGE, BUT IT IS
SURROUNDED BY GENTLE SLOPES.
DESPITE ITS ENORMOUS HEIGHT,
AT THE END OF THESE SLOPES ARE
CLIFFS THAT ARE 7 KM HIGH IN
SOME PLACES.
01010010110
DOWNLOAD COMPLETED
10101100110
CONTINUE
Y/N?

<< IMAGE ENHANCE

9UDGRT56...KD77672DARWY73910-PPE>>>66371020102847201SWYHF7463921

MARTIAN VOLCANOES

INCOMING DATA... ACCEPTED >

MARS IS HOME TO THE LARGEST VOLCANO IN THE SOLAR SYSTEM, AND MANY
OTHER SMALLER ONES. NO ONE KNOWS WHETHER THE MARTIAN VOLCANOES ARE
STILL ACTIVE, BUT IN SOME PLACES THEY SEEM TO HAVE ERUPTED IN THE
LAST FEW MILLION YEARS, WHICH IS QUITE RECENT IN GEOLOGICAL TIME.

> MIGHTY CALDERA

The central pit, or caldera, of Olympus Mons is large enough to swallow most Earth volcanoes whole. It is 52 km across and surrounded by cliffs 6 km deep. The five separate and distinct rings inside the caldera mark different times when the summit collapsed following eruptions from the reservoir of magma beneath it. The youngest ring is about 140 million years old, but lava continued to erupt through fissures on the volcano's sides until about 20 million years ago.

→ DEPTH: 6 KM
WIDTH: 52 KM
MISSION: MARS EXPRESS

01010010110
DOWNLOAD
COMPLETED
10101100110
CONTINUE
Y/N?
10101100101

> THARSIS REGION

Olympus Mons lies on top of a huge bulge in the Martian crust called the Tharsis Rise that helps boost the volcano's height by about 10 km. To the southeast of Olympus Mons lies a chain of three other major volcanoes known as the Tharsis Montes. The Tharsis Rise might be the result of a "hot spot" in the planet's mantle that is pushing the crust upwards, or it could simply be an area where the crust has been thickened by lava that has erupted onto it over billions of years.

→ 1. OLYMPUS MONS
2. ARSIA MONS
3. PAVONIS MONS
4. ASCRAEUS MONS
MISSION: MARS
GLOBAL SURVEYOR
01010010110
DOWNLOAD
COMPLETED
10101100110
CONTINUE
Y/N?

NAME: VALLES
MARINERIS

TYPE: CANYON SYSTEM

DIMENSIONS: 4,000 KM LONG
UP TO 700 KM WIDE

NOTES: INDIVIDUAL CANYONS ARE KNOWN
AS CHASMATA (SINGULAR: CHASMA).
THE TWO LONG CANYONS WEST AND EAST
OF THE CENTRAL BASIN REGION ARE
IUS CHASMA AND COPRATES CHASMA.
010100110110

DOWNLOAD COMPLETED
10101100110

CONTINUE
Y/N?

EXPAND SEARCH>>

DGRT56...KD77672DARWY73910-PPE>>663710201028 47

VALLES MARINERIS

INCOMING DATA... ACCEPTED >

THE MOST IMPRESSIVE FEATURE ON MARS IS THE FAMOUS VALLES
MARINERIS CANYON SYSTEM, NAMED AFTER THE MARINER 9 PROBE
THAT DISCOVERED IT IN 1971. THE "MARINER VALLEYS" ARE TEN
TIMES AS LONG AS EARTH'S GRAND CANYON, UP TO 700 KM ACROSS
AND AN AVERAGE OF 8 KM DEEP.

<< EXPAND SEARCH

> AUREUM CHAOS

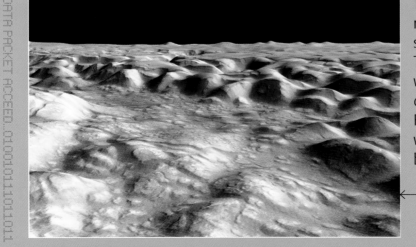

DATA PACKET ACCEID.01001011011011

This area of landslips at the eastern end of the Valles Marineris
reveals much of the canyon's history. It began as a narrow
split in the crust, triggered by pressure from the neighbouring
Tharsis Rise. Water later ran through the valleys, eroding and
widening them and forming sedimentary rocks. When Mars
dried out about 3 billion years ago, wind continued the erosion.
Finally, the disappearance of supporting material (perhaps
water ice) from below ground triggered massive landslides that
broadened the individual valleys.

NAME: AUREUM CHAOS
LOCATION: EASTERN VALLES MARINERIS
DIMENSIONS: A FEW KILOMETRES
MISSION: MARS EXPRESS

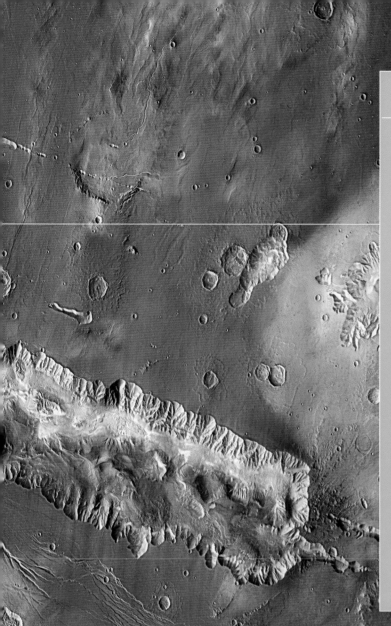

> OPHIR CHASMA

This deep basin lies on the northern side of the enormous series of linked basins that mark the centre of the canyon complex. As with the Aureum Chaos, the central basins probably formed when massive landslips destroyed the higher ground in between individual canyons. Now just a few isolated mesas stand in the midst of the deep basins, though elsewhere in the Valles Marineris, east and west of the basins, long, thin ridges of the original surface still survive.

NAME: OPHIR CHASMA
LOCATION: NORTHERN VALLES MARINERIS
CLIFF HEIGHTS: UP TO 5 KM
MISSION: MARS EXPRESS

DATA PACKET ACCEED .01.001.01111.01.1010011110101010101010101001011011001101

EXPAND SEARCH>>

ASTEROID BELT

BETWEEN MARS AND JUPITER, WE CROSS THE ASTEROID BELT. IT IS HOME TO BILLIONS OF SMALL, ROCKY WORLDS AND A FEW MORE SUBSTANTIAL ONES. DESPITE THE HUGE NUMBER OF OBJECTS OUT THERE, THE REGION OF SPACE THEY OCCUPY IS SO VAST THAT COLLISIONS ARE RARE.

→ ASTEROID 433
PERIHELION: 9.4 LIGHT MINUTES
APHELION: 14.8 LIGHT MINUTES
YEAR LENGTH: 1.76 EARTH YEARS
LENGTH: 31 KM

This Near Earth Asteroid is 31 km long and roughly shaped like a potato. It spends much of its time around the orbit of Mars and inside the main asteroid belt. The asteroid is solid, rocky and peppered with craters left by smaller impacts. Asteroids like Eros have probably not changed much in the 4.6 billion years of the solar system's existence.

> EROS
<< DATA FILE

→ ASTEROID 1
DISTANCE FROM SUN: 22.1 LIGHT MINUTES
YEAR LENGTH: 4.6 EARTH YEARS
DIAMETER: 975 KM

The largest of the asteroids, Ceres is officially a "dwarf planet". Its surface is covered in minerals that formed in the presence of water, and there may be a layer of ice beneath the outer crust. Ceres may also have a thin atmosphere.

> CERES
<< DATA FILE

ASTEROID 4
DISTANCE FROM SUN: 19.6 LIGHT MINUTES
YEAR LENGTH: 3.63 EARTH YEARS
DIAMETER: 530 KM

The second-largest asteroid, Vesta, is about 500 km across and almost spherical except for the enormous impact crater at its south pole. Vesta is also one of a handful of rare asteroids whose surfaces seem to have been changed by volcanic eruptions – something that is hard to understand in a world this size.

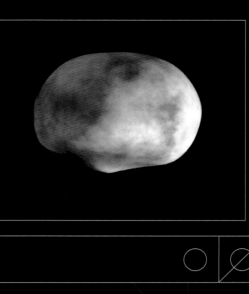

⟫ VESTA
≪ DATA FILE

ASTEROID 243
DISTANCE FROM SUN: 23.8 LIGHT MINUTES
YEAR LENGTH: 4.84 EARTH YEARS
LENGTH OF IDA: 60 KM

Most asteroids follow lone orbits, but some have companions. Ida has a moon called Dactyl, just 1.6 km across, and many other asteroid moons are now suspected. They may be chunks of rock chipped off during impacts. Binary asteroids – evenly matched pairs of worlds – are also quite common.

⟫ IDA AND DACTYL
≪ DATA FILE

ASTEROID 951
DISTANCE FROM SUN: 18.4 LIGHT MINUTES
YEAR LENGTH: 3.29 EARTH YEARS
DIAMETER: 18 KM

This asteroid is rich in silicate rocks similar to those in the crusts of rocky planets. It has relatively few craters on its surface, suggesting that it has not been exposed to space for as long as some other asteroids. Perhaps Gaspra is a fragment from a larger body that broke up.

⟫ GASPRA
≪ DATA FILE

mantle

liquid hydrogen/
helium outer
mantle

atmosphere

distance from Earth: 33 light minutes
diameter: 142,984 km mass: 318 times Earth
day length: 9.9 hours year length: 11.88 Earth years
axial tilt: 3.12° cloudtop gravity: 2.53 g

JUPITER

INCOMING DATA... ACCEPTED >

LARGE ENOUGH TO SWALLOW ALL THE OTHER PLANETS
WITH ROOM TO SPARE, JUPITER IS THE NEAREST GAS
GIANT. IT IS A BLOATED BALL OF HYDROGEN AND
HELIUM, RACKED WITH STORMS AND COLOURED BY THE
CHEMICAL SOUP OF ITS UPPER ATMOSPHERE.

**mean cloudtop
temperature:** −110°C

hazard analysis:
dangerous radiation be
around planet

atmosphere analysis:
90% hydrogen
10% helium

Ganymede

number of moons: 63

RINGS

All the giant planets of the outer solar system have their own ring systems. Saturn's are the most famous, while Jupiter's are little more than a disc of fine dust. The rings are only visible when lit from behind by the Sun, as in this view of Jupiter captured by the Galileo space probe. The dust probably comes from Jupiter's small inner moons and is blasted into orbit by the impact of tiny meteorites.

Jupiter's main ring is roughly 7,000 km wide from its inner to its outer edge but is just 30 km across.

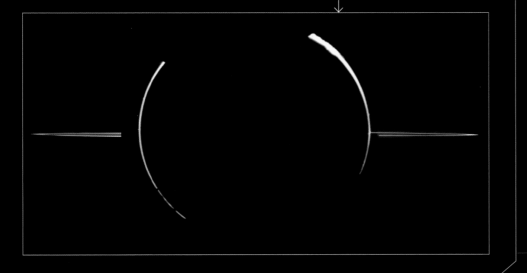

As particles are channelled down onto Jupiter's poles, they collide with atmospheric gases, releasing energy to create glowing aurorae.

Deep inside Jupiter, the pressure is intense enough to break molecules of liquid hydrogen apart into individual atoms. Atomic hydrogen acts like liquid metal, carrying huge electric currents that generate a vast magnetic field around the planet. Jupiter's magnetism creates fierce doughnut-shaped belts of radiation close to the planet and deflects the solar wind, funnelling some of it down to form aurorae at both poles. Its influence is felt as far away as the orbit of Saturn.

MAGNETIC FIELD

SURFACE DETAIL
COMET IMPACT

>> In 1994, Earthbound astronomers had a ringside seat for a spectacular cosmic collision as fragments of Comet Shoemaker-Levy 9 plunged into Jupiter's atmosphere.

>> The comet had been torn apart by a close encounter with Jupiter's gravity in 1992, and by 1994 it had broken into a "string of pearls" – 22 separate fragments all following the same orbit.

>> The collisions happened just on Jupiter's far side, and the largest fragment alone released energy equal to 6 million million tonnes of TNT – 750 times the world's entire arsenal of nuclear weapons.

>> As the planet rotated, huge bruises came into view and were photographed by the Hubble Space Telescope. These deep holes in the upper cloud layers were coloured by chemicals dredged up from deeper inside the planet.

NAME:
EQUATORIAL
CLOUD BANDS

TYPE: WEATHER SYSTEM

DIMENSIONS: ABOUT 10,000 KM WIDE

NOTES: CREAM-COLOURED BANDS ARE
CALLED ZONES, WHILE BLUE AND BROWN
REGIONS ARE CALLED BELTS. ZONES ARE
HIGH-ALTITUDE AMMONIA CLOUDS. BELTS
ARE CLEARINGS THAT REVEAL DEEPER
AMMONIUM HYDROSULPHIDE CLOUDS

1010010110
DOWNLOAD COMPLETED
10101100110
CONTINUE
Y/N?

DATA PACKET ACCEPTED: 010010111011101010001110101010101010101010011

<< IMAGE ENHANCE

> GREAT RED SPOT

Jupiter's most famous feature is the Great Red Spot, a huge area of high pressure roughly twice the size of Earth. It circles the planet in between the dark-coloured South Equatorial Belt and the lighter South Temperate Zone. This enormous storm is coloured by chemicals, drawn up from lower in the atmosphere, which condense into clouds in the cold, high-altitude air. The spot rotates once every seven days, and though its colour sometimes fades, it has survived on Jupiter for at least two centuries.

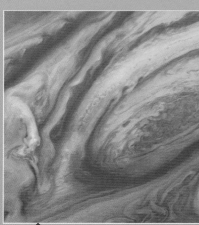

NAME: GREAT RED SPOT
LOCATION: LATITUDE 22° SOUTH
DIMENSIONS: 24,000-40,000 KM BY
12,000-14,000 KM
MISSION: VOYAGER 2

01010010110
DOWNLOAD COMPLETED

JUPITER CLOUDS

INCOMING DATA... ACCEPTED >

THE MULTICOLOURED CLOUDS OF JUPITER FORM BANDS THAT STRETCH AROUND THE PLANET, OFTEN WITH HUGE STORMS BETWEEN THEM. JUPITER'S WEATHER SYSTEMS ARE GOVERNED BY HUGE FORCES, CREATED AS JUPITER COMPLETES A FULL ROTATION IN LESS THAN 10 HOURS. THESE WRAP HIGH-SPEED WINDS ALL THE WAY AROUND THE PLANET.

672DARWY7001029... K0778720RWY73810

<< EXPAND SEARCH

DATA PACKET ACCEPTED.0100101101010100011010100101001011101010

This false-colour picture of Jupiter's cloud bands at mid-northern latitudes reveals hidden features. Thin, high clouds are shown in light blue, thick clouds at mid-altitudes are white, and deeper clouds are red. The purple areas show where a thin haze overlays clear, deep canyons in the atmosphere. Several white storms can clearly be seen swirling between the different bands, and complex patterns called "festoons" run along the boundaries.

NAME: JUPITER
NORTHERN HEMISPHERE
CLOUDS
LOCATION: LATITUDE
10-50° NORTH
DIMENSIONS: APPROX
50,000 KM
MISSION: GALILEO

01010010110
DOWNLOAD COMPLETED
10101100110
CONTINUE
Y/N?
10101100101
10101100101

> CLOUD BANDS

<< IMAGE ENHANCE

metallic core
silicate mantle
sulphurous crust

IO

INCOMING DATA... ACCEPTED >

JUPITER'S INNERMOST MAJOR MOON, IO, IS STAINED
WITH THE YELLOWS, BROWNS AND REDS OF DIFFERENT
FORMS OF SULPHUR. POWERFUL TIDES FROM NEARBY
JUPITER HELP TO HEAT ITS INTERIOR, TURNING IO
INTO THE MOST VOLCANIC BODY IN THE SOLAR SYSTEM.

distance from Jupiter: 421,500 km **mass:** 0.0149 times Earth **orbital period:** 1.769 Earth days
diameter: 3,643 km
rotation period: 1.769 Earth days **axial tilt:** 0° **surface gravity:** 0.183 g

**mean surface
temperature:** −143°C

atmosphere analysis:
90% sulphur dioxide

hazard analysis:
radiation from Jupiter,
volcanoes

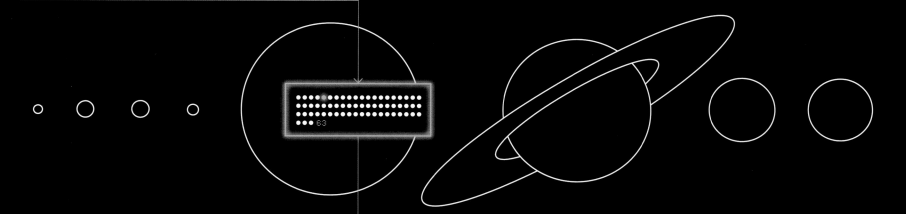

63

> SULPHUR PLUMES

No one suspected Io might be an active world until the Voyager 1 space probe flew past Jupiter and its moons in 1979. As it turned its cameras to look back at Io, Voyager photographed a huge cloud of particles hanging over the moon's horizon. The particles were sulphur compunds condensing from a fountain of liquid sulphur erupting from the surface.

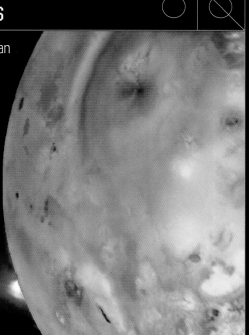

Loki plume, photographed in 1979 by Voyager 1.

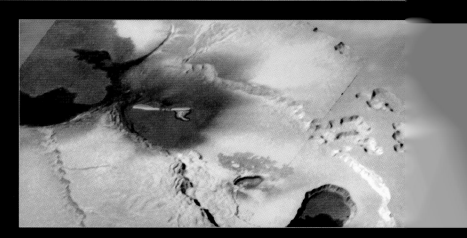

The Galileo space probe took a closer look at Io in the 1990s. It found enormous pools of molten sulphurous rock on the moon's surface marking places where the crust melted and collapsed to form volcani craters or calderas, such as this one in the Tvashtar Catena region.

> VOLCANO WORLD

SURFACE DETAIL
MOON IN TURMOIL

>> As Io orbits Jupiter, the giant planet's gravity pushes and pulls at it, flexing the rocks and generating friction that keeps Io's interior hot.

>> Plumes form where hot reservoirs of molten sulphur, trapped beneath the surface, find a weak point through which to escape. Most of the sulphur in the plumes falls back onto Io as frost, but some escapes the moon's gravity and ends up orbiting Jupiter.

>> Continuous eruptions from plumes and calderas rapidly transform the surface, redrawing the map of Io in just a few decades.

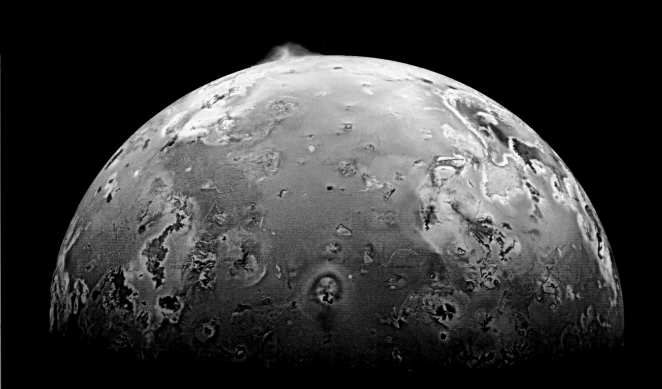

EUROPA

INCOMING DATA... ACCEPTED >

THE SECOND OF JUPITER'S MAJOR MOONS IS ALSO THE SMALLEST, BUT THIS ICE-COVERED BALL OF ROCK CONCEALS ONE OF THE BIGGEST SECRETS IN THE SOLAR SYSTEM. IT HAS DEEP OCEANS OF LIQUID WATER THAT MAKE IT A POSSIBLE HAVEN FOR ALIEN LIFE.

metallic

silicate mantle

sub-surface ocean

icy crust

distance from Jupiter: 671,000 km

mass: 0.0083 times Earth

orbital period: 3.55 Earth days

diameter: 3,138 km

rotation period: 3.55 Earth days

axial tilt: 0°

surface gravity: 0.135 g

mean surface temperature: −148°C

atmosphere analysis:
trace oxygen

hazard analysis:
radiation from Jupiter and the Sun

44

63

<< IMAGE ENHANCE

> SURFACE CRACKS

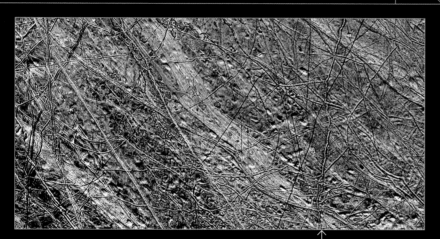

Europa's surface is covered with brown and reddish streaks, thought to form when tides from Jupiter cause the crust to shift and crack. Water stained with sulphur and other chemicals wells up from the ocean below. Europa's sea freezes rapidly on exposure to the cold, almost airless surface, leaving lasting scars across the surface of the moon.

Individual streaks are flanked by ridges on either side.

This field of fractured ice on Europa resembles pack ice seen on Earth, but in reality the frozen crust of Europa is far thicker. Lines running in various directions show where the surface has split apart and re-frozen many times in the past. The bright, blue-white material is debris from the young crater Pwyll.

Older ice floes are coloured by mineral-stained water.

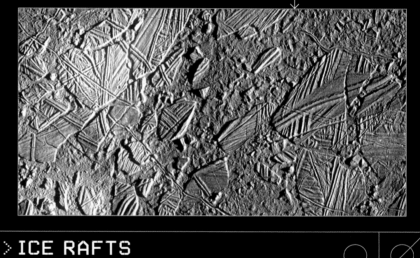

> ICE RAFTS
<< IMAGE ENHANCE

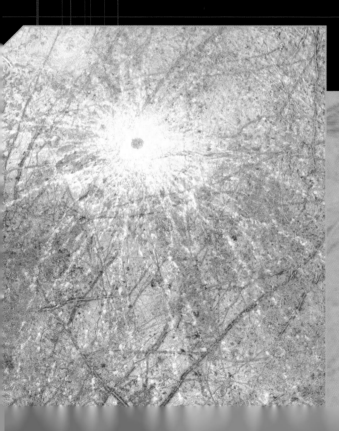

SURFACE DETAIL
SELF-HEALING MOON

>> Beneath a crust of ice several kilometres thick, Europa conceals a global ocean that is kept warm by heat from the interior. The moon takes a similar tidal pounding to its neighbour Io, and almost certainly has undersea volcanoes belching chemicals into the depths. On Earth, similar "black smokers" are ideal habitats for deep-sea creatures.

>> The hidden ocean and constantly shifting ice crust help keep the moon uniquely smooth. For its size, Europa is smoother than a cueball. The result can be seen in the fate of Europa's craters – Pwyll (left) is young and fresh, while Tyre (right), a larger but older impact basin, has flattened out and slumped until it is little more than a "stain".

GANYMEDE

INCOMING DATA... ACCEPTED >

JUPITER'S THIRD MAJOR
SATELLITE IS THE LARGEST MOON
IN THE SOLAR SYSTEM. GANYMEDE
LIES FAR ENOUGH FROM JUPITER
TO NOT SUFFER THE SAME TIDAL
HEATING AS IO OR EUROPA, BUT
ITS COMPLEX SURFACE REVEALS A
PAST THAT WAS MUCH MORE ACTIVE.

partially molten
core
rocky mantle

liquid water
ocean

icy crust

distance from Jupiter: 1.007 million km
diameter: 5,262 km mass: 0.0247 times Earth
rotation period: 7.15 Earth days orbital period: 7.15 Earth days
axial tilt: 0° surface gravity: 0.145 g

atmosphere analysis:
trace hydrogen
ozone

mean surface temp:
-164°C

hazard analysis:
radiation from Jupiter
and the Sun

SURFACE DETAIL
GROOVED LANDSCAPES

>> Ganymede's terrain is a mix of dark and light patches,
often separated by bands of long, parallel grooves.
The surface is icy, the interior is rocky and there are hints
of a layer of liquid water between the two.

>> These complex features suggest this world was once
warmed by the same tidal heating that still affects its inner
neighbours. A combination of the gravity of the other
moons probably once put Ganymede in an orbit that came
much closer to Jupiter.

>> The jumbled terrain and parallel grooves show how the
ancient surface partially melted, with some parts sinking
down into the interior and others pulling apart, rather like
Earth's own tectonic plates.

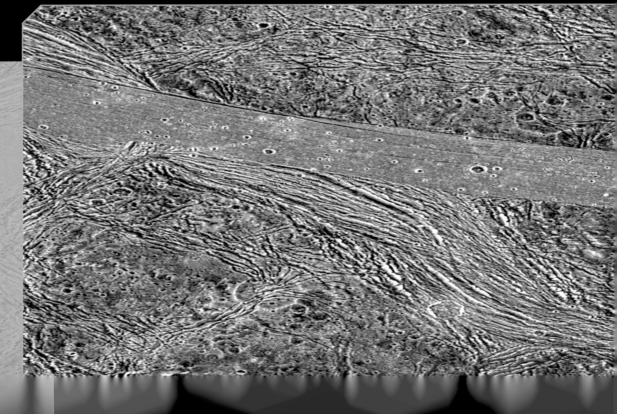

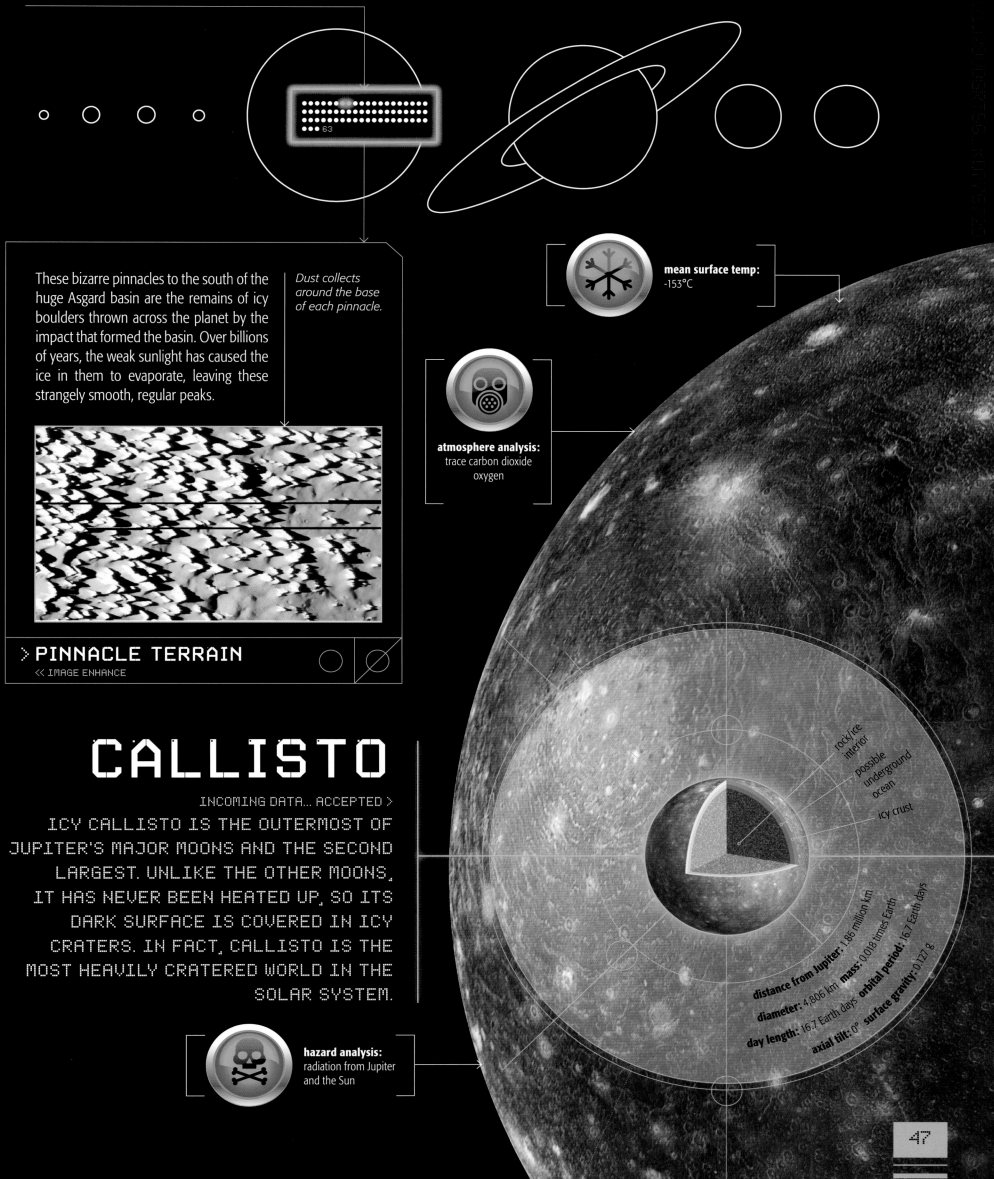

These bizarre pinnacles to the south of the huge Asgard basin are the remains of icy boulders thrown across the planet by the impact that formed the basin. Over billions of years, the weak sunlight has caused the ice in them to evaporate, leaving these strangely smooth, regular peaks.

Dust collects around the base of each pinnacle.

> PINNACLE TERRAIN
<< IMAGE ENHANCE

mean surface temp:
-153°C

atmosphere analysis:
trace carbon dioxide
oxygen

CALLISTO

INCOMING DATA... ACCEPTED >

ICY CALLISTO IS THE OUTERMOST OF JUPITER'S MAJOR MOONS AND THE SECOND LARGEST. UNLIKE THE OTHER MOONS, IT HAS NEVER BEEN HEATED UP, SO ITS DARK SURFACE IS COVERED IN ICY CRATERS. IN FACT, CALLISTO IS THE MOST HEAVILY CRATERED WORLD IN THE SOLAR SYSTEM.

hazard analysis:
radiation from Jupiter
and the Sun

rock/ice interior

possible underground ocean

icy crust

distance from Jupiter: 1.86 million km
diameter: 4,806 km **mass:** 0.018 times Earth
day length: 16.7 Earth days **orbital period:** 16.7 Earth days
axial tilt: 0° **surface gravity:** 0.127 g

47

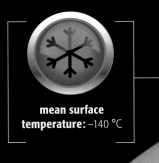

rock/ice core
liquid metallic hydrogen inner mantle
liquid hydrogen /helium outer mantle
atmosphere

distance from Earth: 71 light minutes **mass:** 95 times Earth
diameter: 120,536 km **year length:** 29.46 Earth years
day length: 10.66 hours **cloudtop gravity:** 1.07 g
axial tilt: 26.7°

SATURN

INCOMING DATA... ACCEPTED >

FAMOUS FOR ITS RINGS, SATURN IS THE SECOND
LARGEST WORLD IN THE SOLAR SYSTEM. THIS
BALL OF GAS AND LIQUID IS SIMILAR TO
JUPITER, BUT IT IS LESS DENSE THAN WATER,
SO IT WOULD FLOAT IN A LARGE ENOUGH OCEAN.

mean surface temperature: −140 °C

« IMAGE ENHANCE
> BLUE LIGHT

Saturn's sepia tone is mainly due to a haze of ammonia that
forms in the high, cold air of the upper atmosphere. The rings
cast complex shadows onto the clouds, but they can also give
much of the planet an eerie blue tint. Sunlight passing through
fine particles in the rings is scattered in the same way as when
it passes through Earth's atmosphere, turning it blue.

atmosphere analysis:
93% hydrogen
7% helium

Hyperion

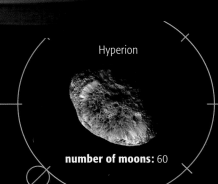

number of moons: 60

Changing ultraviolet aurorae from 24–28 January 2004.

Although Saturn's magnetic field is not as powerful as Jupiter's, it is still strong enough to sweep up particles from the solar wind and generate bright aurorae at the planet's magnetic poles – as seen in these images of the changing lights around the south polar region.

> ## POLAR LIGHTS
<< IMAGE ENHANCE

hazard analysis:
high-speed particles
travelling in ring plane

SURFACE DETAIL
STORMY ATMOSPHERE

>> Saturn's hazy upper atmosphere gives the planet a deceptively calm appearance. In reality Saturn is just as complex and stormy as Jupiter, as revealed in this false-colour image (right).

>> The picture, from the Cassini probe, shows a huge electrical storm raging on Saturn. High atmospheric clouds are shown in white, while lower layers appear red. The root of the storm lies deep below these visible cloud layers.

>> Elsewhere on the planet, oval white storms come and go. While Saturn has no equal to the Great Red Spot, large "Great White Spots" develop every 30 years or so around midsummer for the northern hemisphere.

SATURN'S RINGS

INCOMING DATA... ACCEPTED >

SATURN'S SPECTACULAR RINGS ARE MADE OF COUNTLESS RINGLETS, EACH CONTAINING BILLIONS OF SEPARATE FRAGMENTS OF ICY MATERIAL. DIFFERENCES IN THE THICKNESS AND COMPOSITION OF THE RINGS, AND THE EFFECT OF GRAVITY FROM NEARBY MOONS, DIVIDE THEM INTO JUST A FEW DISTINCT SECTIONS: THE FAINT D RING CLOSEST TO THE PLANET, THE TRANSLUCENT C RING, THE CREAMY-WHITE B RING, AND THE OUTER A AND F RINGS.

MAJOR FEATURES:
D RING:
6,632 TO 14,242 KM FROM SATURN
C RING:
14,390 TO 31,732 KM FROM SATURN
B RING:
31,732 TO 57,312 KM FROM SATURN
CASSINI DIVISION
INNER A RING: 61,902 TO 73,159 KM FROM SATURN ENCKE DIVISION
OUTER A RING: 73,484 TO 76,507 KM FROM SATURN

> D RING

> C RING

≪ IMAGE ENHANCE
> CHANGING VIEWS

DATA PACKET ACCEPTED.010

Saturn's tilted axis means that from our point of view on Earth, we see the rings from different angles throughout the Saturnian year. Sometimes they are "wide open", so we can see across them clearly. At other times they are edge-on to us, and because they are only a few kilometres thick, they disappear through all but the most powerful telescopes.

1996   1997   1998   1999

DATA PACKET ACCEPTED..01000101

<< IMAGE ENHANCE

> RING RAINBOW

All the giant planets have ring systems, but only Saturn's are so grand. This is partly because they contain so much material, but also because they are largely made from frozen water ice and so reflect more sunlight than rings of other material. This "rainbow" recorded by the Cassini probe is a result of sunlight being reflected back off a point directly opposite the Sun as the fast-moving probe took images of the rings through three coloured filters.

Particle sizes in the A ring are up to 100 m in diameter.

73910-RFE>%63710201102847201911NHF746392

> A RING

> B RING

<< IMAGE ENHANCE

> DETAILED STRUCTURE
<< IMAGE ENHANCE

→ *Tiny moonlets orbiting within the rings create propeller-shaped gaps around themselves.*

Up close, the rings break down into countless ringlets, resembling the grooves in a vinyl record. Each ringlet is made from a stream of particles following a very tight circular orbit around Saturn. Any particles that stray from their path are soon knocked back into line, or thrown out of the system altogether, by collisions with their neighbours. Gravity from the outer moons and also from "shepherd moons" that orbit between the rings helps clear distinct gaps, such as the Cassini Division. Smaller moonlets, a few kilometres across, replenish the rings with new material knocked off them in repeated impacts.

→ *The moonlets in the rings may be tens of metres across. Particles can also group together into clumps a few kilometres wide.*

DATA PACKET ACCEPTED..01101010101010010001100011000110001011101

INNER MOONS

INCOMING DATA... ACCEPTED >

SATURN'S COMPLEX SYSTEM OF MOONS HAS NO RIVAL IN THE SOLAR SYSTEM. ALONGSIDE THE USUAL COLLECTION OF TINY SHEPHERD MOONS AMONG THE RINGS, AND MORE REMOTE, CAPTURED ASTEROIDS, THERE ARE MORE THAN A DOZEN UNIQUE, FASCINATING WORLDS.

PANDORA
<< DATA FILE

AVERAGE DISTANCE FROM SATURN: 141,700 KM
ORBITAL PERIOD: 15.1 HOURS
DIAMETER: 86 KM
SURFACE TEMPERATURE: -200°C
SURFACE GRAVITY: 0.001 G

10101100101 10101100101

Pandora and its twin Prometheus are shepherd moons that orbit on either side of the narrow outer F ring – their gravity keeps the ring particles in line. Material thrown off the moons in meteorite impacts replenishes the ring, which is constantly losing material that drifts down towards Saturn.

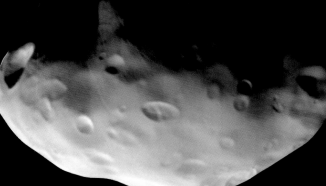

MIMAS
<< DATA FILE

AVERAGE DISTANCE FROM SATURN: 185,500 KM
ORBITAL PERIOD: 22.6 HOURS
DIAMETER: 418 KM
SURFACE TEMPERATURE: -200°C
SURFACE GRAVITY: 0.007 G

10101100101 10101100101

The closest of the large moons to Saturn is Mimas. This heavily cratered, icy world is dominated by the Herschel Crater, which at 140 km across is one-third of the diameter of the entire moon. It makes Mimas resemble the Death Star from the *Star Wars* films.

EPIMETHEUS
<< DATA FILE

AVERAGE DISTANCE FROM SATURN: 151,500 KM
ORBITAL PERIOD: 16.7 HOURS
DIAMETER: 122 KM
SURFACE TEMPERATURE: -195°C
SURFACE GRAVITY: 0.001 G

This irregularly shaped moon is one of a pair with the slightly larger Janus. These "co-orbital" moons follow almost identical orbits around Saturn, swapping positions every four years. Epimetheus itself has had huge chunks knocked off it by impacts in the past, leaving it with a series of flattened "facets".

AVERAGE DISTANCE FROM SATURN:
238,050 KM
ORBITAL PERIOD: 32.9 HOURS
DIAMETER: 512 KM
SURFACE TEMPERATURE: -198°C
SURFACE GRAVITY: 0.012 G

A little larger and farther out than Mimas, Enceladus is an unexpectedly active world. Its surface is brilliant white, making it the most reflective world in the solar system. Strong tidal forces heat its interior, so that liquid water runs just beneath the surface and escapes into space as plumes of ice.

> ENCELADUS
<< DATA FILE

SURFACE DETAIL
ICE PLUMES ON ENCELADUS

>> The jets of ice above Enceladus were only confirmed in 2005 when the Cassini space probe flew through them during a close flyby of the moon. They form around warmer patches in the icy crust (marked by blue "tiger stripes" in the main, colour-enhanced image), where water running under the surface can find a weak point and boil away into space.

>> Fresh snow falls back onto Enceladus and rapidly covers any craters that form there, leaving a landscape of pristine snow.

MIDDLE MOONS

INCOMING DATA... ACCEPTED >

FARTHER OUT FROM SATURN LIE THREE MID-SIZED MOONS: TETHYS, DIONE AND RHEA. TETHYS AND DIONE ARE ABOUT THE SAME SIZE AND SHOW SIGNS OF ACTIVITY IN THEIR ANCIENT PAST. RHEA IS LARGER THAN ITS NEIGHBOURS BUT STRANGELY HAS FEWER SIGNS OF ACTIVITY.

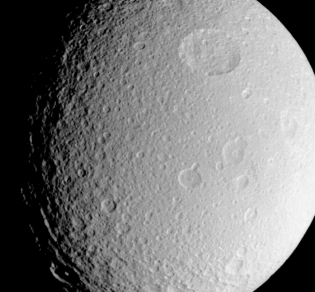

> TETHYS
<< DATA FILE

AVERAGE DISTANCE FROM SATURN:
294,700 KM
ORBITAL PERIOD: 45.3 HOURS
DIAMETER: 1,058 KM
SURFACE TEMPERATURE: -187°C
SURFACE GRAVITY: 0.015 G

Looking like a larger version of Enceladus, Tethys has a fairly bright surface but no signs of recent activity. Its larger craters have slumped around their edges, suggesting that the moon's crust is made up of an ammonia/water ice mix that can flow like the ice in glaciers on Earth.

AVERAGE DISTANCE FROM SATURN:
527,000 KM
ORBITAL PERIOD: 108.5 HOURS
DIAMETER: 1,530 KM
SURFACE TEMPERATURE: -197°C
SURFACE GRAVITY: 0.028 G

The countless craters on Rhea suggest that it has been deep-frozen since it formed, with no activity to smooth away craters as they accumulated. Scientists think that Rhea's higher gravity caused it to freeze into a rare, rock-hard form of ice that could not flow like that on Tethys or Dione. Rhea is the only satellite in the solar system with its own ring of orbiting material.

> RHEA
<< DATA FILE

AVERAGE DISTANCE FROM SATURN:
294,700 KM
ORBITAL PERIOD: 45.3 HOURS
DIAMETER: 24 KM
SURFACE TEMPERATURE: -190°C
SURFACE GRAVITY: UNKNOWN

Two small moons, Telesto and Calypso, share an orbit with Tethys. Telesto is oddly smooth for such a tiny world. It may be blanketed with icy particles that the moon sweeps up from Saturn's E ring, a broad ring of fine material through which the inner moons orbit.

> TELESTO
<< DATA FILE

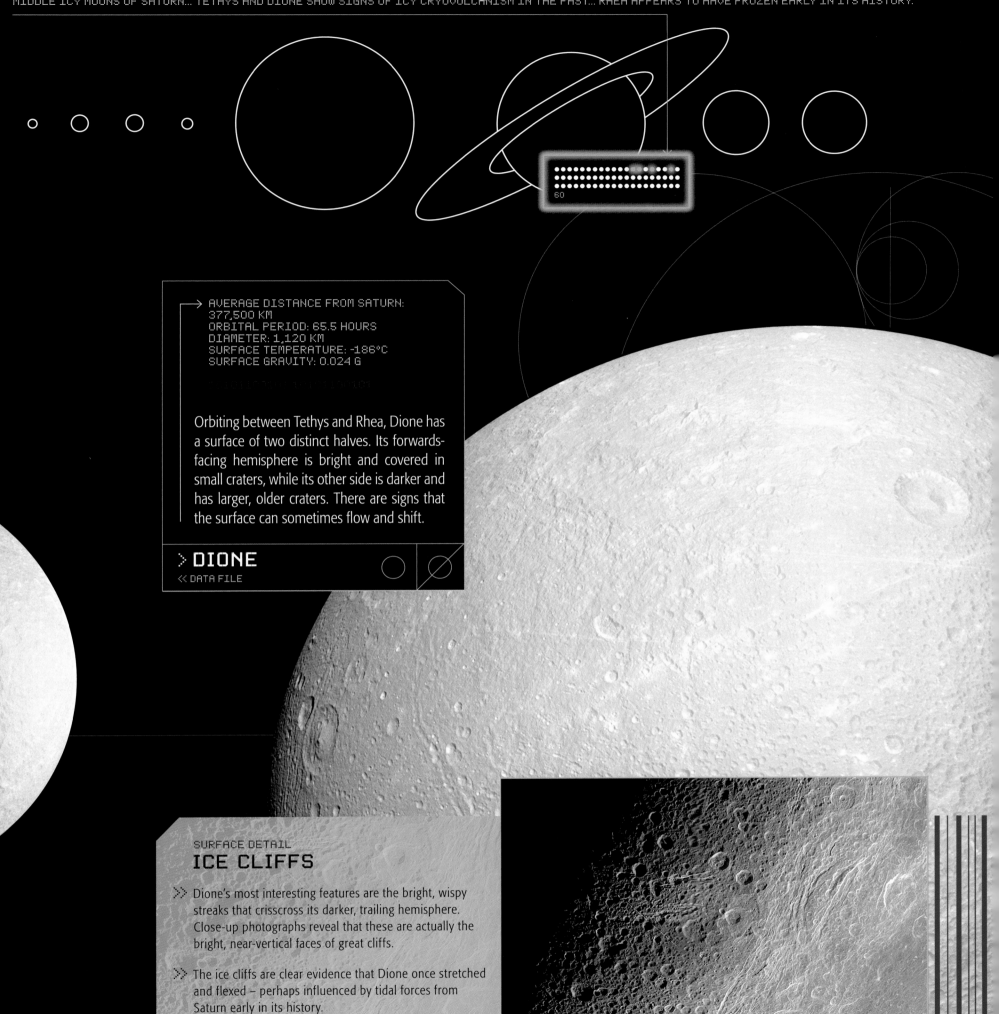

AVERAGE DISTANCE FROM SATURN:
377,500 KM
ORBITAL PERIOD: 65.5 HOURS
DIAMETER: 1,120 KM
SURFACE TEMPERATURE: -186°C
SURFACE GRAVITY: 0.024 G

Orbiting between Tethys and Rhea, Dione has a surface of two distinct halves. Its forwards-facing hemisphere is bright and covered in small craters, while its other side is darker and has larger, older craters. There are signs that the surface can sometimes flow and shift.

›› DIONE
‹‹ DATA FILE

SURFACE DETAIL
ICE CLIFFS

›› Dione's most interesting features are the bright, wispy streaks that crisscross its darker, trailing hemisphere. Close-up photographs reveal that these are actually the bright, near-vertical faces of great cliffs.

›› The ice cliffs are clear evidence that Dione once stretched and flexed – perhaps influenced by tidal forces from Saturn early in its history.

›› The same tides may have triggered "cryovolcanism" on the bright hemisphere, allowing fresh ammonia and ice to ooze onto the surface and wipe out all traces of the older, larger craters.

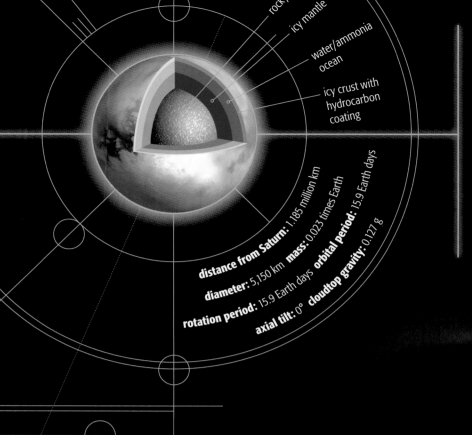

rocky
icy mantle
water/ammonia ocean
icy crust with hydrocarbon coating

distance from Saturn: 1.185 million km
diameter: 5,150 km mass: 0.023 times Earth
rotation period: 15.9 Earth days orbital period: 15.9 Earth days
axial tilt: 0° cloudtop gravity: 0.127 g

TITAN

INCOMING DATA... ACCEPTED >

TITAN IS SATURN'S LARGEST SATELLITE AND THE ONLY ONE OF THE SOLAR SYSTEM'S MOONS WITH A SUBSTANTIAL ATMOSPHERE. THIS OPAQUE LAYER OF ORANGE-COLOURED GASES HIDES A STRANGELY EARTH-LIKE SURFACE DOMINATED AND SHAPED BY THE SOLID, LIQUID AND GASEOUS FORMS OF METHANE.

mean surface temp:
-179°C

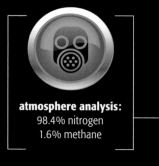

atmosphere analysis:
98.4% nitrogen
1.6% methane

hazard analysis:
atmosphere of combustible methane

<< IMAGE ENHANCE

BENEATH THE CLOUDS

Titan's atmosphere, like Earth's, is dominated by nitrogen gas, but methane clouds and a layer of orange smog hide the surface from view. In 2004, the Cassini space probe arrived in orbit around Saturn, carrying with it an infrared camera able to see through the clouds for the first time. Titan turned out to have an icy crust similar to Saturn's other moons, except that it has very few craters and has been shaped by the forces of erosion.

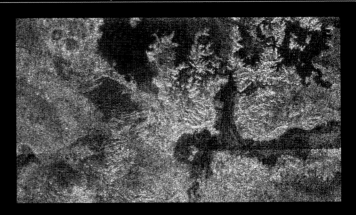

In 2006, Cassini detected the first signs of liquid on the surface of Titan – lakes of methane around the north pole. On the freezing surface of Titan, methane seems to play the same role as water does on Earth, freezing into ice, evaporating into vapour and falling back to the surface as rain.

> METHANE LAKES
<< IMAGE ENHANCE

SURFACE DETAIL

DESCENT TO THE SURFACE

>> In January 2005, Cassini flew overhead as the Huygens lander parachuted into Titan's atmosphere and sent back the first direct pictures of the surface. As the probe emerged from the cloud, it photographed an Earth-like landscape – apparently a coastline with rolling hills and river deltas.

>> Huygens landed on a dried-out lakebed covered in a thin crust of ice. Photographs from the surface showed it surrounded by icy rocks carried into the lake from a nearby river. Instruments indicated that although the landing site was dry, it had probably rained quite recently.

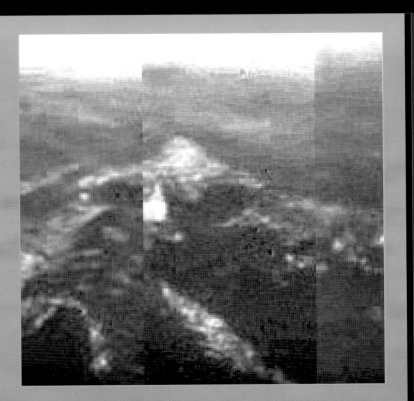

OUTER MOONS

INCOMING DATA... ACCEPTED >

BEYOND TITAN LIE THREE OTHER MAJOR SATURNIAN MOONS, EACH BIZARRE AND UNUSUAL IN ITS OWN WAY. THEY ARE THE STRANGE, MISSHAPEN HYPERION, WITH A VARIABLE AND UNPREDICTABLE ROTATION, THE RHEA-SIZED IAPETUS, WITH ITS BIZARRE SURFACE MARKINGS, AND THE DARK, IRREGULAR PHOEBE.

> HYPERION
<< DATA FILE

AVERAGE DISTANCE FROM SATURN:
1,481,000 KM
ORBITAL PERIOD: VARIABLE (CHAOTIC)
LENGTH: 370 KM
SURFACE TEMPERATURE: -190°C
SURFACE GRAVITY: 0.002 G

Sponge-like Hyperion is a bizarre world that scientists don't fully understand. It probably got its odd appearance as sunlight melted surface ice to leave "spines" of rocky material. This moon is probably just a fragment of a larger world that broke up in a collision.

> IAPETUS
<< DATA FILE

AVERAGE DISTANCE FROM SATURN:
3,561,250 KM
ORBITAL PERIOD: 79.3 EARTH DAYS
DIAMETER: 1,440 KM
SURFACE TEMPERATURE: -190°C
SURFACE GRAVITY: 0.028 G

This moon has one hemisphere that is icy and bright and another that is covered in unidentified dark material. There seem to be traces of this material on some other Saturnian moons, but we still don't know where it comes from.

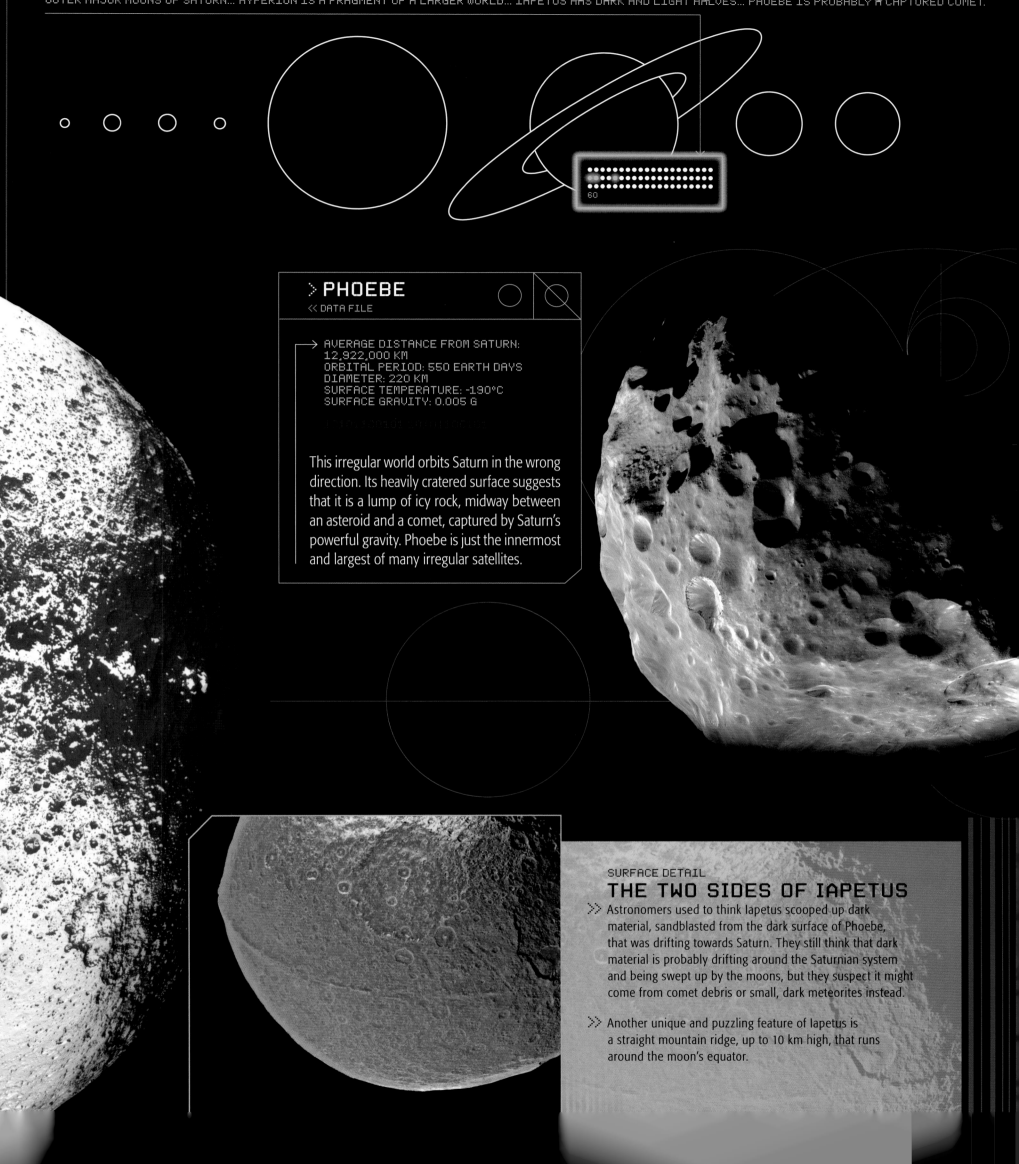

60

>> PHOEBE
<< DATA FILE

→ AVERAGE DISTANCE FROM SATURN:
12,922,000 KM
ORBITAL PERIOD: 550 EARTH DAYS
DIAMETER: 220 KM
SURFACE TEMPERATURE: -190°C
SURFACE GRAVITY: 0.005 G

This irregular world orbits Saturn in the wrong direction. Its heavily cratered surface suggests that it is a lump of icy rock, midway between an asteroid and a comet, captured by Saturn's powerful gravity. Phoebe is just the innermost and largest of many irregular satellites.

SURFACE DETAIL
THE TWO SIDES OF IAPETUS

>> Astronomers used to think Iapetus scooped up dark material, sandblasted from the dark surface of Phoebe, that was drifting towards Saturn. They still think that dark material is probably drifting around the Saturnian system and being swept up by the moons, but they suspect it might come from comet debris or small, dark meteorites instead.

>> Another unique and puzzling feature of Iapetus is a straight mountain ridge, up to 10 km high, that runs around the moon's equator.

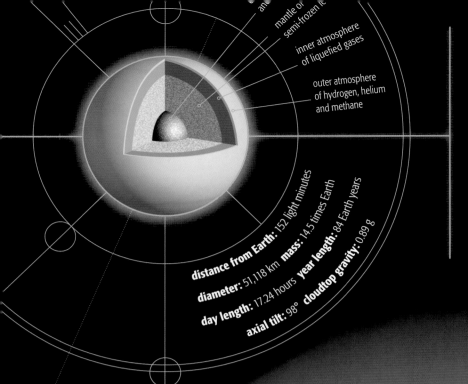

mantle or
semi-frozen ice

inner atmosphere
of liquefied gases

outer atmosphere
of hydrogen, helium
and methane

distance from Earth: 152 light minutes

diameter: 51,118 km **mass:** 14.5 times Earth

day length: 17.24 hours **year length:** 84 Earth years

axial tilt: 98° **cloudtop gravity:** 0.89 g

URANUS

INCOMING DATA... ACCEPTED >

URANUS IS THE FIRST OF THE SOLAR SYSTEM'S TWO
"ICE GIANTS", PLANETS WHOSE OUTER ATMOSPHERES
COVER A SEA OF SLUSHY ICE. UNIQUELY, ITS AXIS OF
ROTATION IS TILTED AT MORE THAN 90 DEGREES FROM
VERTICAL, GIVING IT A BIZARRE CYCLE OF SEASONS.

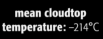

**mean cloudtop
temperature:** −214°C

hazard analysis:
dark and difficult
to detect rings

atmosphere analysis:
83% hydrogen
15% helium
2.3% methane

Oberon

number of moons: 27

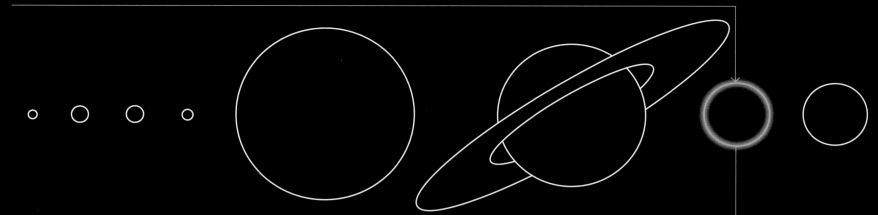

<< IMAGE ENHANCE

RINGS OF URANUS

Uranus is encircled by 11 narrow rings, mostly made of frozen methane ice. The rings are so dark that they were only discovered in 1977 when they passed in front of a star and caused its light to flicker. Material in the rings is herded into narrow paths, between 1 km and 13 km across and less than 15 km deep, by the influence of small shepherd moons that orbit either side of the rings.

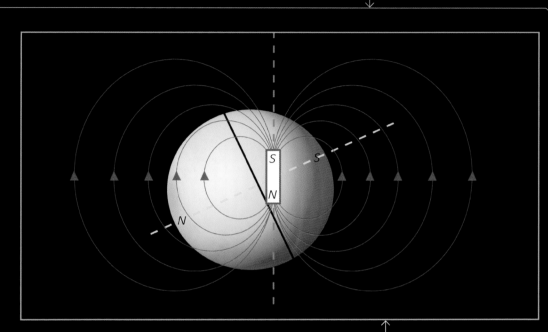

Uranus was probably knocked sideways by a huge collision early in its history. For half of each 84-year orbit, the north pole is in daylight, and for the other half, the south pole gets the sunlight. For much of each orbit, the strong difference in temperature stops weather systems, such as bands and storms, from developing.

Uranus's magnetic field has an even stranger alignment than its rotation – it does not even pass through the centre of the planet.

> WORLD OFF ITS AXIS
<< IMAGE ENHANCE

SATELLITE INFORMATION

MOONS OF URANUS

>> Uranus has five major moons, each shaped in a different way by its own internal heat and tidal forces. Two of them, Titania (left) and Miranda (right), are shown here.

>> Tiny Miranda has a jumbled surface that shows it suffered extreme heating from the tides early in its history – enough to melt it almost completely and rearrange the surface into a chaotic jigsaw.

>> Ariel has had less heating from the tides, but still enough to trigger icy cryovolcanism, wiping out craters and opening up deep trenches. Umbriel, farther out and the same size, is dark and heavily cratered.

>> Titania looks like a larger version of Ariel. The energy for its cryovolcanoes came from heat left over from its formation rather than tidal forces. Oberon is similar, but because it is slightly smaller, it has seen less activity and so more craters survive.

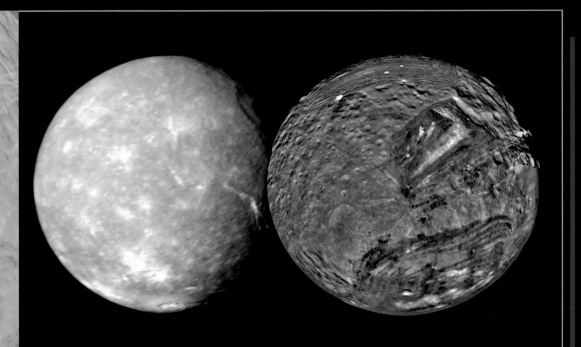

NEPTUNE

INCOMING DATA... ACCEPTED >

THE OUTERMOST MAJOR PLANET, DEEP-BLUE NEPTUNE IS ANOTHER ICE GIANT LIKE URANUS. IT IS ALSO A WORLD OF HIGH WINDS AND FIERCE STORMS, POWERED BY ENERGY FROM INSIDE THE PLANET AS WELL AS BY HEAT FROM THE DISTANT SUN.

mantle or
semi-frozen
inner atmosphere
of liquefied gases

outer atmosphere
of hydrogen, helium
and methane

distance from Earth: 248 light minutes

diameter: 49,532 km **mass:** 17.1 times Earth

day length: 16.11 hours **year length:** 164.9 Earth years

axial tilt: 28.3° **cloudtop gravity:** 1.13 g

mean cloudtop temperature: −200°C

hazard analysis: extreme high winds in atmosphere

atmosphere analysis:
80% hydrogen
19% helium
1.5% methane

Nereid

number of moons: 13

› WORLD OF STORMS

Neptune gets its colour from the methane in its atmosphere, which absorbs red light. (Uranus has slightly more methane and so is blue-green in colour.) Storms in the atmosphere form dark spots that come and go every few years. Occasionally, weather systems can wrap themselves around the planet to form bands similar to those on the inner giants. Neptune has an Earth-like pattern of seasons, but this far out, spring lasts for more than 40 years.

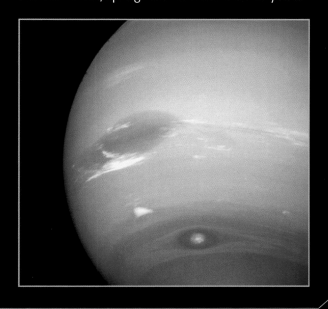

White clouds high in the atmosphere are stretched into long streams by Neptune's winds, which are some of the fastest in the solar system at up to 2,000 kph. Most of the energy for Neptune's powerful weather comes from inside the planet, where chemical changes deep beneath the surface release heat. One theory is that the main change is caused by carbon atoms combining into crystals – if this is correct, then deep inside Neptune, it literally rains diamonds. Jupiter and Saturn also have internal power plants, but Neptune's is the strongest.

› WHITE STREAMERS

SATELLITE INFORMATION
MOONS OF NEPTUNE

>> Neptune has a few small shepherd moons around its three narrow rings, and a few other small moons farther out, but the entire system is dominated by Triton, a single large moon. Triton orbits in a perfect circle but is going the "wrong way" around the planet.

>> Triton is probably an ice dwarf world from the Kuiper Belt (see pages 64–65), captured into orbit after a close encounter with Neptune. Strong tides changed the shape of its orbit into a perfect circle and warmed up the interior enough to trigger cryovolcanism and create geysers that still belch dust and gas from the surface.

>> Most of the original moons that got in Triton's way were flung out of the system altogether, but one – Nereid – seems to have clung on in a highly elongated, tilted orbit.

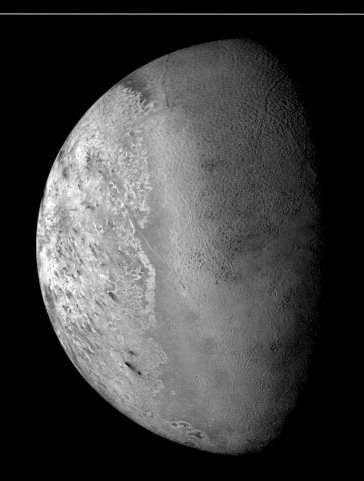

rock/ice

mantle – mostly
water ice

crust of
various ices

perihelion: 4.1 light hours **aphelion:** 6.8 light hours
diameter: 2,304 km **mass:** 0.02 times Earth
day length: 6.4 Earth days **year length:** 248.6 Earth years
axial tilt: 119.6° **cloudtop gravity:** 0.06 g

PLUTO, THE KUIPER BELT AND BEYOND

INCOMING DATA... ACCEPTED >
ONCE CONSIDERED A PLANET, PLUTO IS IN FACT
AN ICE DWARF. IT IS PART OF THE KUIPER BELT OF
ICY WORLDS THAT ORBIT THE SUN JUST BEYOND
NEPTUNE. PLUTO HAS THREE MOONS, THE LARGEST OF
WHICH, CHARON, IS MORE THAN HALF ITS SIZE.

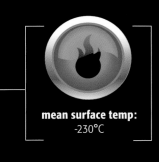

mean surface temp:
-230°C

atmosphere analysis:
trace nitrogen,
methane

hazard analysis:
extreme cold,
icy surface

> ERIS
<< DATA FILE

In 2003, astronomers discovered Eris, an object in the Kuiper Belt that was slightly larger than Pluto. This forced them to redefine what qualifies as a planet. Now Pluto, Eris and the asteroid Ceres are classed as "dwarf planets" – big enough to be spherical but without enough gravity to have cleared other objects out of their orbits. Eris is at least 2,400 km across and takes 560 years to orbit the Sun.

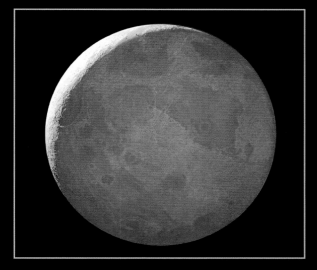

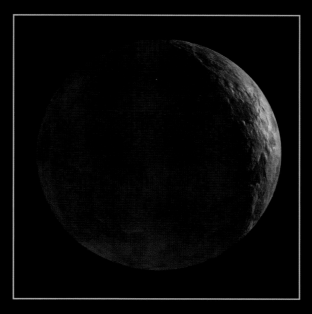

Sedna, the most distant object discovered so far in the solar system, was found in 2003. It is also the coldest world, with a surface temperature of -240°C, and takes at least 12,000 years to circle the Sun in a highly elongated orbit. Sedna lies far beyond the Kuiper Belt and may be an inner member of the Oort Cloud. However, at 1,500 km across, it could also be an escaped dwarf planet from the inner solar system. It is the second reddest object in the solar system after Mars.

> SEDNA
<< DATA FILE

DEBRIS CLOUDS
KUIPER BELT AND OORT CLOUD

>> Two great clouds of small, icy bodies encircle the solar system beyond the orbit of Neptune – the doughnut-shaped Kuiper Belt and the spherical Oort Cloud.

>> Kuiper Belt objects are the icy equivalents of asteroids. Pluto and Eris are the largest so far discovered, but there are probably many more of the same size. Many short-period comets also linger here on the outer edges of their orbits.

>> The Oort Cloud is home to trillions of deep-frozen comet nuclei and is the source of long-period comets that take thousands of years to circle the Sun. The cloud stretches out to one light year from the Sun, and occasional disturbances and collisions send new comets falling towards the inner solar system.

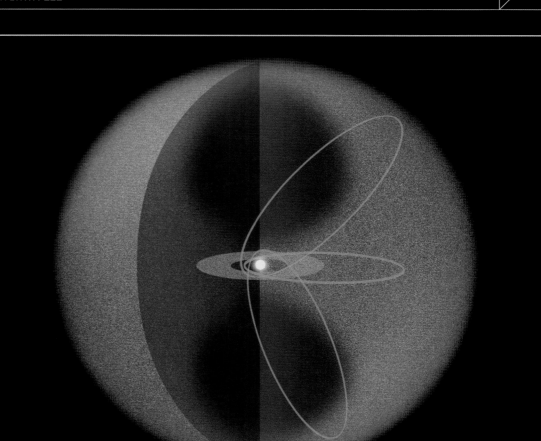

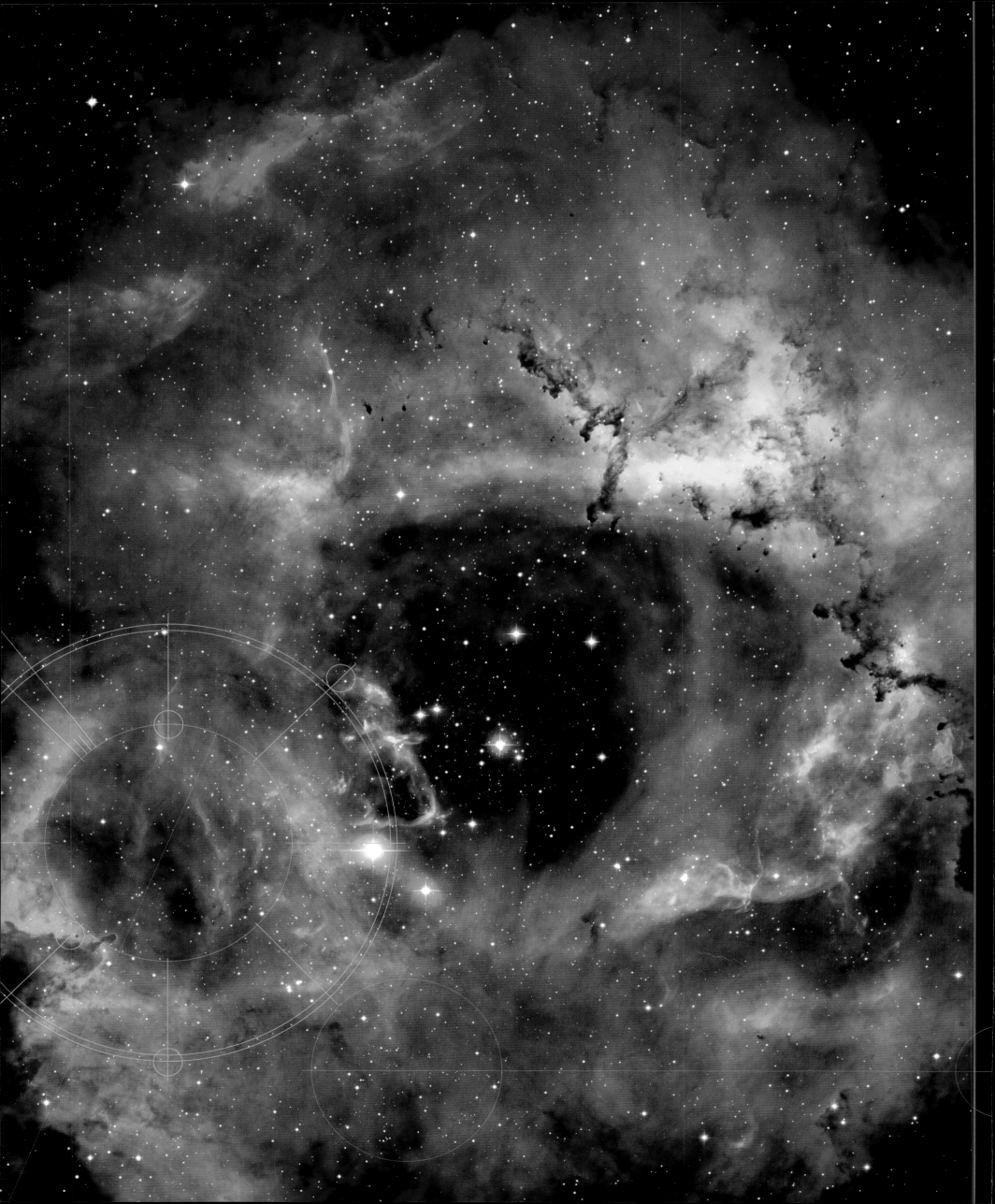

THROUGH THE MILKY WAY

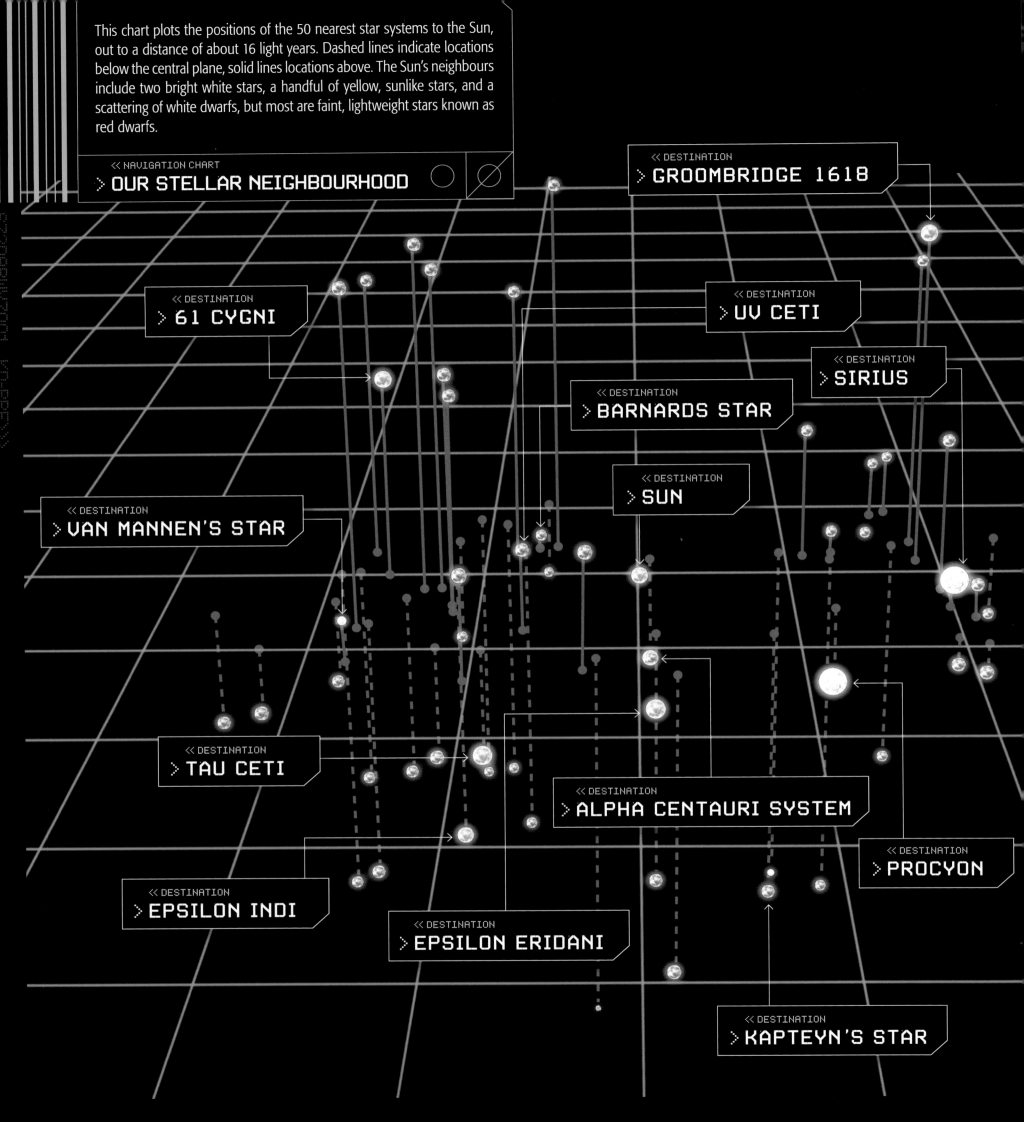

This chart plots the positions of the 50 nearest star systems to the Sun, out to a distance of about 16 light years. Dashed lines indicate locations below the central plane, solid lines locations above. The Sun's neighbours include two bright white stars, a handful of yellow, sunlike stars, and a scattering of white dwarfs, but most are faint, lightweight stars known as red dwarfs.

« NAVIGATION CHART
> OUR STELLAR NEIGHBOURHOOD

« DESTINATION
> GROOMBRIDGE 1618

« DESTINATION
> 61 CYGNI

« DESTINATION
> UV CETI

« DESTINATION
> SIRIUS

« DESTINATION
> BARNARDS STAR

« DESTINATION
> SUN

« DESTINATION
> VAN MANNEN'S STAR

« DESTINATION
> TAU CETI

« DESTINATION
> ALPHA CENTAURI SYSTEM

« DESTINATION
> PROCYON

« DESTINATION
> EPSILON INDI

« DESTINATION
> EPSILON ERIDANI

« DESTINATION
> KAPTEYN'S STAR

AMONG THE STARS

Our Sun is just one among 200 billion stars in our local galaxy, the Milky Way. Almost every object visible in the night sky is in fact a Sun in its own right, and together they, with the clouds of dust and gas that float among them, reveal the story of stellar life and death.

Stars are born from huge clouds of gas that collapse and pull themselves together through gravity until their central regions become hot and dense enough to ignite and begin the nuclear reactions that power them. As the gas and dust clear, stars emerge in loose clusters, but many form in tighter orbits around each other, creating binary pairs and multiples.

All stars spend most of their lives shining by nuclear fusion, converting hydrogen, the lightest and most plentiful element in the Universe, into helium, the next lightest element. The more massive a star is, the denser its core and the fiercer its nuclear fire – the heaviest stars can be tens of thousands of times more luminous than the Sun, while the lightest can be thousands of times dimmer. The Sun lies somewhere in the middle of the brightness and size range, which is just as well for us – its sedate lifestyle means it won't run out of fuel for another five billion years (about the same length of time that it's already shone). More massive stars might shine brilliantly, but they can squander their energy in just a few million years.

However, all stars eventually run out of hydrogen in their cores, and the fusion process starts to run down. Dying stars pass through a series of transformations that form some of the strangest and most spectacular objects in the sky. Changes to their internal structure as the core runs down trigger nuclear reactions in different layers of the star, which cause it to brighten by a factor of hundreds or thousands of times, while its surface expands and cools, usually turning red. This "red giant" phase can last for millions of years, and comes and goes as the star finds new material to burn in the core.

The final fate of a star depends on its mass. In stars like the Sun, the red giant eventually becomes unstable and puffs off its outer layers into a complex, ghostly shell known as a planetary nebula. Because the nuclear reactions effectively hold the star up against its own gravity, once they are gone the core slowly collapses into a small, dense and intensely hot object about the size of the Earth – a white dwarf.

If a star has more than eight times the mass of the Sun, however, it will carry on shining up until the last moment. When the nuclear reactions in the core cut off, the star suddenly collapses inwards, then rebounds off the core in a huge explosion – a supernova. The collapsed core that remains is much smaller and denser than a white dwarf – it forms a neutron star or even a black hole.

And yet the death of stars also marks the beginning of new life – planetary nebulae and supernovae scatter elements across space to enrich new star-forming nebulae, and the shockwave from a supernova can even trigger new generations of starbirth – ultimately leading to planets, rocks and even living creatures that are eventually made from the remains of long-dead stars.

NAME: KEYHOLE
NEBULA
DESIGNATION: NGC 3372

DISTANCE: 8,000 LIGHT YEARS
DIAMETER: 300 LIGHT YEARS

NOTABLE FEATURES:
LARGE STARBIRTH NEBULA SHAPED BY
RADIATION FROM YOUNG STARS. HOST TO
THE ETA CARINAE NEBULA (SEE PAGE 92).

01010010110
DOWNLOAD COMPLETED
10101100110
CONTINUE
Y/N?

INCOMING DATA... ACCEPTED >

STARS ARE BORN IN ENORMOUS COLLAPSING CLOUDS OF GAS AND DUST CALLED NEBULAE. BRIGHT PARTS OF THE CLOUDS SHINE BY REFLECTING STARLIGHT AND BY EMISSION, GLOWING AS THEIR GAS ATOMS ARE EXCITED BY RADIATION FROM NEWBORN STARS. DARK AND DUSTY AREAS FORM HUGE COLUMNS THAT ENGULF STARS AND BLOCK OUT LIGHT FROM THE BACKGROUND. WITHIN THEM, INDIVIDUAL DARK CLOUDS, CALLED BOK GLOBULES, COLLAPSE UNDER THEIR OWN GRAVITY, PULLING IN MORE MATERIAL FROM THEIR SURROUNDINGS.

⟩ HORSEHEAD NEBULA

The most famous of all dark nebulae is the Horsehead Nebula, a towering column of gas about 16 light years across, which contains the same mass of material as 16 Suns. The nebula is sculpted into its elegant chess-piece shape by fierce radiation from the nearby star Sigma Orionis. It is silhouetted against the glowing gas of a more distant emission nebula, IC 434. Parallel streaks of light in the emission nebula are probably created by gas flowing along a weak magnetic field.

DESIGNATION: BARNARD 33
DIAMETER: 16 LIGHT YEARS
DISTANCE: 1,500 LIGHT YEARS

01010010110
DOWNLOAD COMPLETED
10101100110

⟩ LAGOON NEBULA

At the heart of the Lagoon Nebula, radiation and particles streaming from newborn stars, such as the bright, superhot Herschel 36 (lower right), have transformed dark gas clouds into a pair of cosmic tornadoes. The Lagoon is one of the brightest and largest starbirth nebulae in Earth's skies – it has already given rise to several generations of young stars. It is still actively forming stars today inside dark globules that emerge as the rest of the dust is blown away.

DESIGNATION: M8, NGC 6523
DIAMETER: 70 LIGHT YEARS
DISTANCE: 5,000 LIGHT YEARS

01010010110
DOWNLOAD COMPLETED
10101100110
CONTINUE
Y/N?

NAME: EAGLE NEBULA
CLASSIFICATION: IC 4703

DIAMETER (TOTAL):
70 LIGHT YEARS
DISTANCE: 6,500 LIGHT YEARS

NOTABLE FEATURES:
THE NEBULA'S BRIGHTEST STARS
ARE MEMBERS OF M16, A RECENTLY
FORMED STAR CLUSTER THAT MEASURES
ABOUT 15 LIGHT YEARS ACROSS.

01010010110
DOWNLOAD COMPLETED
10101100110
CONTINUE
Y/N?

<< EXPAND SEARCH

> PILLARS OF CREATION

These are huge towers of gas that are enriched with dust from giant, dark columns inside the nebula. Each pillar is several light years long. The pillars are gradually separating into ball-shaped "Bok globules", each measuring a light year or more across. These globules are the beginnings of solar systems – as they continue to collapse under their own gravity, their centres will eventually ignite as new stars.

→ DISTANCE:
6,500 LIGHT YEARS
LENGTH:
9.5 LIGHT YEARS
01010010110
DOWNLOAD
COMPLETED
10101100110
CONTINUE
Y/N?
10101100101
10101100101

<< EXPAND SEARCH

Gas and dust in the nebula is delicately sculpted by particles and fierce radiation blowing away from the new-born stars of the M16 cluster. The nebula is lit by a mix of reflected light from nearby stars and fluorescence as the gas is excited by invisible ultraviolet radiation. The radiation also heats up the gas, allowing it to escape from the pillar's gravity and boil into surrounding space, as seen around the edges of this dark dust cloud.

→ DISTANCE:
6,500 LIGHT
YEARS
LENGTH:
1 LIGHT YEAR
01010010110
DOWNLOAD
COMPLETED
10101100110
CONTINUE
Y/N?
10101100101
10101100101

> GLOWING CLOUDS

EAGLE NEBULA

INCOMING DATA... ACCEPTED >

THE EAGLE NEBULA IN THE SAGITTARIUS ARM OF THE MILKY WAY IS ONE OF THE MOST FAMOUS AND COMPLEX STAR-FORMING NEBULAE. IT IS AN ENORMOUS REGION OF COLLAPSING GAS AND DUST THAT IS ILLUMINATED FROM WITHIN BY THE FIERCE RADIATION OF YOUNG STARS BORN ABOUT 5 MILLION YEARS AGO. THE NEBULA IS SURROUNDED BY DARKER DUST CLOUDS THAT HOLD IT IN A ROUGHLY SPHERICAL REGION ABOUT 70 LIGHT YEARS ACROSS.

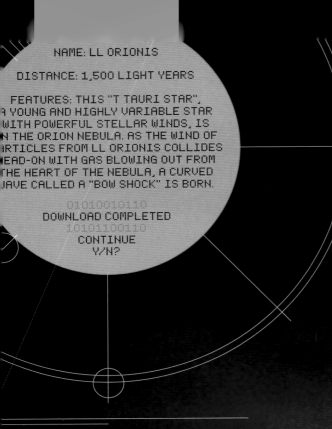

NAME: LL ORIONIS

DISTANCE: 1,500 LIGHT YEARS

FEATURES: THIS "T TAURI STAR",
A YOUNG AND HIGHLY VARIABLE STAR
WITH POWERFUL STELLAR WINDS, IS
IN THE ORION NEBULA. AS THE WIND OF
PARTICLES FROM LL ORIONIS COLLIDES
HEAD-ON WITH GAS BLOWING OUT FROM
THE HEART OF THE NEBULA, A CURVED
WAVE CALLED A "BOW SHOCK" IS BORN.

01010010110
DOWNLOAD COMPLETED
10101100110
CONTINUE
Y/N?

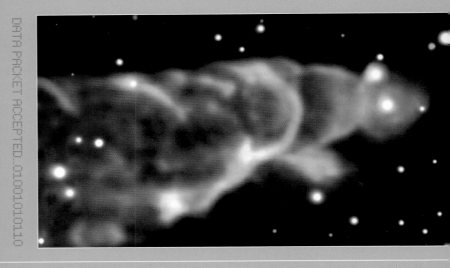

DATA PACKET ACCEPTED..01001010110

<< EXPAND SEARCH
> HH 49/50

<< EXPAND SEARCH
> NGC 1999

DATA PACKET ACCEPTED..01001011101001101111010010101010101001001001110101010101010100100110010101010

This ghostly nebula in Orion emits
no light of its own – it shines only by
reflecting light from the young star,
V380 Orionis, close to its centre. The
star is in turn almost eclipsed by the
dark, T-shaped cloud in front of it –
a Bok globule. This clump of cold,
dense gas and dust probably contains
at least one collapsing protostar
within it.

DIAMETER: 0.9
LIGHT YEARS
DISTANCE:
1,500 LIGHT
YEARS

01010010110
DOWNLOAD
COMPLETED
10101100110
CONTINUE
Y/N?

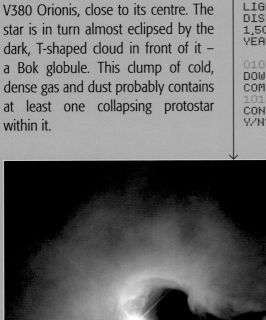

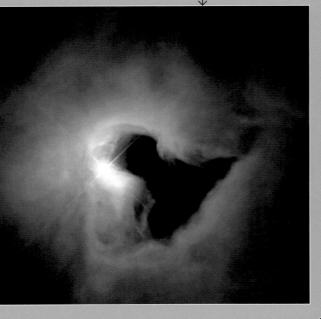

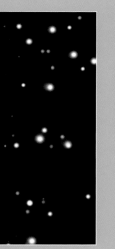

Herbig-Haro object HH 49/50 is a long, V-shaped nebula apparently emerging from the star at the centre of this picture. In reality, an unseen youthful star off the left-hand side of the image is spitting out a high-speed jet that slams into nearby interstellar gas and causes it to glow. The glowing gas spreads out in a "V" shape, just like the wake behind a speedboat. Different colours indicate different amounts of energy released in the collision.

DATA PACKET ACCEPTED: 0100110111011010010010101010101011010101010101010101011010101010

The central star in this picture is suffering stellar growing pains that have caused it to fling jets of excess gas into space from its poles. The jets are invisible as they travel across space, until they plough into the remnants of its original star-forming nebula at some distance from the star. This causes the dust and gas to billow out above and below the star in glowing clouds, known as Herbig-Haro (HH) objects. Nearby, gas glows gently pink as it is excited by the young star's fierce radiation.

DIAMETER: 0.4
LIGHT YEARS
DISTANCE: 1,000
LIGHT YEARS

01010010110
DOWNLOAD
COMPLETED
10101100110
CONTINUE
Y/N?

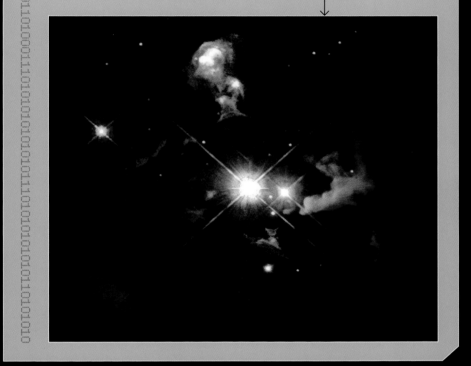

YOUNG STARS

INCOMING DATA... ACCEPTED >

STARS ARE BORN WHEN A COLLAPSING BALL OF GAS WITHIN A NEBULA BECOMES HOT AND DENSE ENOUGH TO TRIGGER NUCLEAR REACTIONS IN ITS CORE. BECAUSE A YOUNG STAR IS STILL SURROUNDED BY MUCH OF THE MATERIAL IN WHICH IT FORMED, IT REMAINS UNSTABLE FOR SOME TIME, OFTEN VARYING IN BRIGHTNESS WITH UNPREDICTABLE AND VIOLENT RESULTS. IT MAY THROW OFF EXCESS MATERIAL AS HIGH-SPEED JETS IN LINE WITH ITS POLES OF ROTATION.

DOUBLES AND CLUSTERS

INCOMING DATA... ACCEPTED >

MORE THAN HALF OF ALL STARS ARE MEMBERS OF BINARY OR MULTIPLE SYSTEMS.
THESE ARE SMALL GROUPS OF STARS BORN FROM THE SAME CLUMP OF GAS AND
FOREVER LOCKED IN ORBIT AROUND EACH OTHER. LARGER GROUPS ARE CALLED
CLUSTERS AND MAY CONTAIN VERY YOUNG OR VERY OLD STARS.

NAME: THE TRAPEZIUM

DIAMETER: 1.5 LIGHT YEARS
DISTANCE: 1,600 LIGHT YEARS

NOTABLE FEATURES:
EACH OF THE FOUR MASSIVE, BRIGHT
YOUNG STARS IN THIS SMALL, TIGHT
CLUSTER SHINES WITH THE LIGHT OF
MORE THAN 10 MILLION SUNS. THIS
BRIGHT QUADRUPLE IS PART OF A MUCH
WIDER CLUSTER IN THE ORION NEBULA
SPREAD ACROSS 20 LIGHT YEARS.

01010010110
DOWNLOAD COMPLETED
10101100110
CONTINUE
Y/N?

DATA PACKET ACCEPTED...010010110010010011010101010010010001110101010100100

<< EXPAND SEARCH

> MIRA

Mira, or Omicron Ceti, is a famous double-star system. Its brighter star, Mira A, is a huge red giant that changes its brightness by a factor of up to 1,000 every 332 days thanks to fluctuations in its size and internal structure. Detailed images of Mira A show a huge plume of material being pulled away from the star towards Mira B, where it forms a disc around the smaller star. Such close relationships are typical of binary stars.

DESIGNATION: OMICRON CETI
LUMINOSITY: UP TO
8,000 SUNS
DISTANCE: 400 LIGHT YEARS

01010010110
DOWNLOAD COMPLETED
10101100110

This huge ball of red and yellow stars is a globular cluster. It contains several hundred thousand small stars, each on average about half the mass of our Sun. Globular clusters orbit close to the centre of the Milky Way and in a "halo" above and below our galaxy. They seem to contain very old stars that formed about the same time as the Milky Way itself and have shone steadily for billions of years. Massive black holes near their centres may help to keep them together.

≪ EXPAND SEARCH

> M80

DESIGNATION: MAMAJEK 1
NUMBER OF STARS: 18
DISTANCE: 316 LIGHT YEARS

01010010110
DOWNLOAD COMPLETED
10101100110
CONTINUE

This loose group of young stars is an open cluster – a stellar family that has only recently emerged from its cocoon of gases. Random motion among its stars will cause the cluster to disintegrate in a few million years, but for now, our view is dominated by young and unpredictable T Tauri stars.

≪ EXPAND SEARCH

> ETA CHAMELEONTIS

NAME: THE PLEIADES
DESIGNATION: M45

DISTANCE: 440 LIGHT YEARS

DIAMETER: 43 LIGHT YEARS

FEATURES: YOUNG OPEN STAR CLUSTER
RICH IN MASSIVE BLUE-WHITE STARS.
OVERALL MASS OF 800 SUNS.

010100101110
101011001110

DOWNLOAD COMPLETED

CONTINUE
Y/N?

<< IMAGE ENHANCE

> RADIATION SIGNATURE ANALYSIS

DATA PACKET ACCEPTED..01001011101011010001

Infrared images of the Pleiades reveal the true extent of the gas and dust still surrounding the young stars. This Spitzer space telescope image shows the densest regions of warm cosmic dust in red and yellow and the outlying regions in green. Because the hot, bright stars of the Pleiades give out most of their light at other wavelengths, they look dimmer in infrared, while other, cooler stars emerge from the gloom.

THE PLEIADES

INCOMING DATA... ACCEPTED >

THE PLEIADES CLUSTER, OR SEVEN SISTERS, IS ONE OF THE MOST FAMOUS OPEN CLUSTERS, A GROUP OF YOUNG BLUE-WHITE STARS THAT FORMED ABOUT 100 MILLION YEARS AGO. THE CLUSTER CONTAINS ABOUT 1,000 STARS IN TOTAL, WITH SEVEN BEING BRIGHT ENOUGH TO SEE FROM EARTH WITH THE NAKED EYE.

<< IMAGE ENHANCE

<< IMAGE ENHANCE

DATA PACKET ACCEPTED..010010111101101100011010101001010010011001

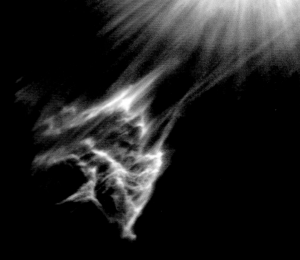

Merope is one of about a dozen bright, blue-white stars that dominate our view of the Pleiades. Most of the stars in the cluster are a lot less massive, so they shine less brightly, but will live much longer. The entire cluster is still surrounded by a tangled web of ghostly gas – the remains of the nebula from which it formed.

<< IMAGE ENHANCE
> MEROPE

NAME: NGC
3603

DISTANCE: 20,000 LIGHT YEARS

NOTABLE FEATURES:
THE LIFE STORY OF STARS CAPTURED
IN A SINGLE IMAGE. THE CENTRAL
STAR CLUSTER HAS RECENTLY EMERGED
FROM A NEBULA, SUCH AS THAT AT LOWER
RIGHT. MEANWHILE, A BRILLIANT BLUE
SUPERGIANT AT TOP LEFT, SURROUNDED
BY A BUBBLE OF GAS, LURCHES
TOWARDS A SUPERNOVA EXPLOSION
THAT WILL SCATTER ITS
MATERIAL BACK
ACROSS SPACE.

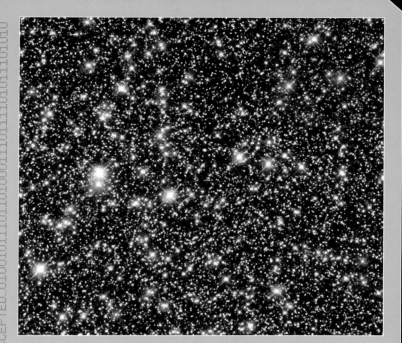

This spectacular star cloud shows how stars vary in colour and luminosity. The colour of a star depends on its surface temperature – orange and red stars are relatively cool, while white and blue stars are much hotter. Blue stars are also more compact than red stars of the same luminosity.

≪ EXPAND SEARCH
> SAGITTARIUS STAR CLOUD

≪ EXPAND SEARCH
> PROXIMA CENTAURI

This speckled image shows the closest star to the Sun, Proxima Centauri, which is a typical red dwarf star. With less than an eighth of the Sun's mass, it shines with just 1/7,000th of its luminosity, and is only visible through powerful telescopes despite a distance of just 4.2 light years. The vast majority of our galaxy's stars are dwarfs like Proxima Centauri.

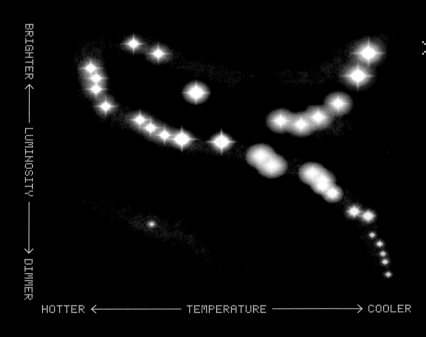

BRIGHTER ← LUMINOSITY → DIMMER

HOTTER ← TEMPERATURE → COOLER

≪ EXPAND SEARCH
> BRIGHTNESS AND TEMPERATURE

This graph, called the Hertzsprung-Russell diagram, is a simple way of seeing the rules that stars obey. It compares the temperature of a star with its luminosity. When enough stars are plotted onto the diagram, nearly all of them turn out to fall along a diagonal stream called the "main sequence", linking the dim red stars with bright blue ones. Stars on this strip also obey another rule – the more massive they are, the brighter they shine. Our Sun lies somewhere near the middle, among the yellow stars classed as type "G". Some stars break the rules: red and yellow giants are bright but cool (so they must have enormous surfaces), while white dwarfs are dim but hot (so they must be tiny).

THE LIVES OF STARS

INCOMING DATA... ACCEPTED >

ONCE A YOUNG STAR HAS SETTLED DOWN, IT SHINES STEADILY FOR MANY MILLIONS, OR EVEN BILLIONS, OF YEARS, POWERED BY NUCLEAR REACTIONS IN ITS CORE. FOR MOST OF ITS LIFE, THE LUMINOSITY (TRUE BRIGHTNESS) AND COLOUR OF A STAR DEPEND ON ITS MASS, OBEYING A RELATIONSHIP

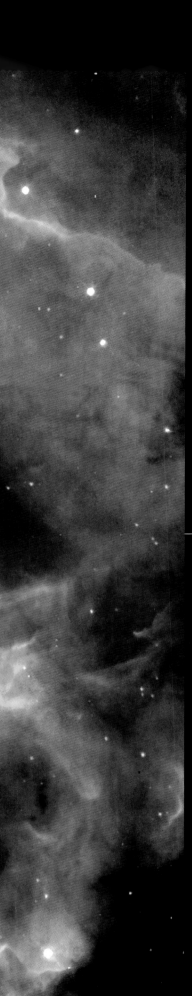

NAME:
PSR 1257+12
TYPE: PULSAR
PLANETARY SYSTEM

DISTANCE: 980 LIGHT YEARS

NOTABLE FEATURES:
THIS BIZARRE SOLAR SYSTEM ORBITING
A BURNT-OUT PULSAR WAS DETECTED
LONG BEFORE ANY OTHERS DUE TO
ITS EFFECT ON THE PULSAR'S RADIO
SIGNALS. IT CONSISTS OF TWO URANUS-
SIZED PLANETS, ONE ABOUT THE SIZE
OF MERCURY AND A SMALL ASTEROID.
01010010110
DOWNLOAD COMPLETED
10101100110
CONTINUE
Y/N?

« EXPAND SEARCH

› FOMALHAUT

DATA PACKET ACCEPTED. 010010111010

One of the brightest stars in Earth's sky, Fomalhaut is a young white star about 250 million years old. Infrared images show it is surrounded by a broad disc of dusty material that probably has planets forming within it. Sometimes referred to as Fomalhaut's own Kuiper Belt, from Earth the disc looks like an eye staring into space. It is also off-centre from its parent star – perhaps due to the influence of a large planet that has already formed close to the star.

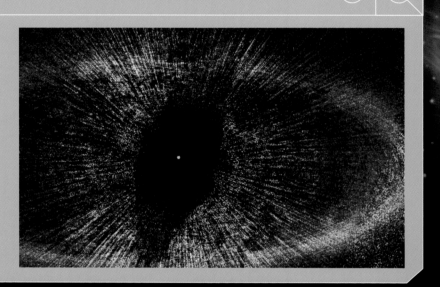

EXTRASOLAR PLANETS

OUR SOLAR SYSTEM IS NOT UNIQUE. MANY OTHER STARS HAVE PLANETS ORBITING THEM OR FORMING AROUND THEM, ALTHOUGH WE CAN ONLY DETECT THESE "EXTRASOLAR" PLANETS IF THEY ARE ABOUT THE SIZE OF JUPITER. OUR SOLAR SYSTEM'S NEAR-CIRCULAR ORBITS, ROCKY INNER PLANETS AND COLD OUTER GAS GIANTS ARE UNUSUALLY REGULAR. OTHER SOLAR SYSTEMS SEEM MORE CHAOTIC.

> HD 107146

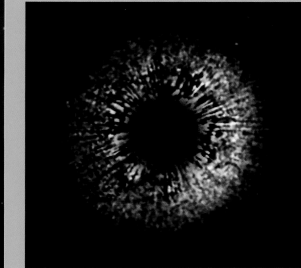

This ring of reddish, dusty material surrounds a recently formed Sun-like star that is a few tens of millions of years old and 88 light years from Earth. The ring sits at about the same distance from its star as the Kuiper Belt does from the Sun, and the orbits of Neptune and all the inner planets would fit neatly into the central gap. However, this dust ring contains at least a thousand times more material than the Kuiper Belt, suggesting that the system will develop in a different way.

DATA PACKET ACCEPTED..010010111011010111

>EXTRASOLAR PLANET 2M1207B

DATA PACKET ACCEPTED..010010111000011111111

The first extrasolar planet to be photographed directly, 2M1207b is a gas giant with the mass of five Jupiters. It orbits a dim and feeble brown dwarf star that itself weighs only four times more than the planet. Brown dwarfs are extremely faint objects, somewhere between small stars and giant planets, that form in the same way as proper stars. The brown dwarf 2M1207 and its planet 2M1207b are both extremely young, and the planet is still glowing red hot – making it easy to detect in this infrared image.

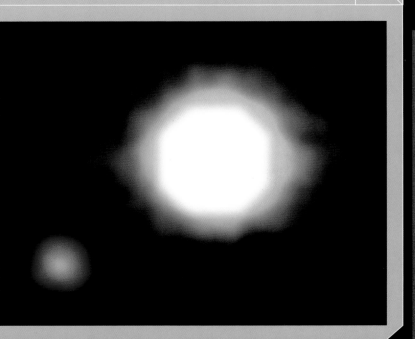

STRANGE STARS

INCOMING DATA... ACCEPTED >

STARS CAN VARY UNPREDICTABLY FOR A NUMBER OF
REASONS, BUT USUALLY BECAUSE OF SOME CHANGE IN
THEIR INTERNAL STRUCTURE OR THE WAY THAT THEY
PRODUCE THEIR ENERGY. IN SOME OF THE MOST VIOLENT
CASES, CALLED NOVAE, THE CHANGE IS CAUSED BY THE WAY
THAT TWO STARS IN A BINARY SYSTEM INTERACT.

<< EXPAND SEARCH

> XZ TAURI

This binary star system in Taurus, which is blowing out a huge bubble of expanding
hot gas, was captured by three images in 1995, 1998 and 2000. Superhot gas from an
unseen disc around one of the stars is squeezed through their crashing magnetic fields
before being channelled outwards at almost 540,000 kph. The expanding bubble emits
light as it begins to cool farther from the source.

DATA PACKET ACCEPTED. 01001011101101011 01001011011011011 010

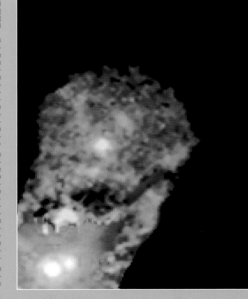

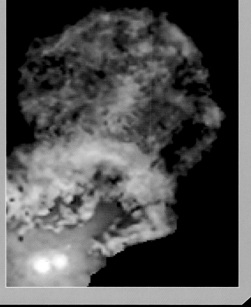

> NOVA CYGNI

DATA PACKET ACCEPTED...01000101101011101

Nova outbursts are explosions on the surface of a dense white dwarf star in a binary system. In Nova Cygni's case, the white dwarf's companion star has swollen into a red giant, pushing its flimsy outer layers into the grasp of the dwarf's powerful gravity. As the dwarf pulls material away, it piles up on its own surface, eventually becoming so dense that it explodes in a burst of nuclear reactions. These two images show the expanding shockwave of Nova Cygni a year and two years after the explosion.

RING DIAMETERS:
119 AND 155 BILLION KM
DISTANCE: 10,400 LIGHT YEARS

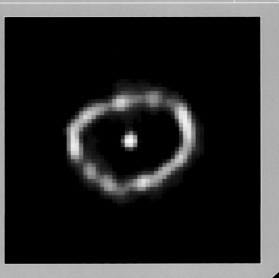

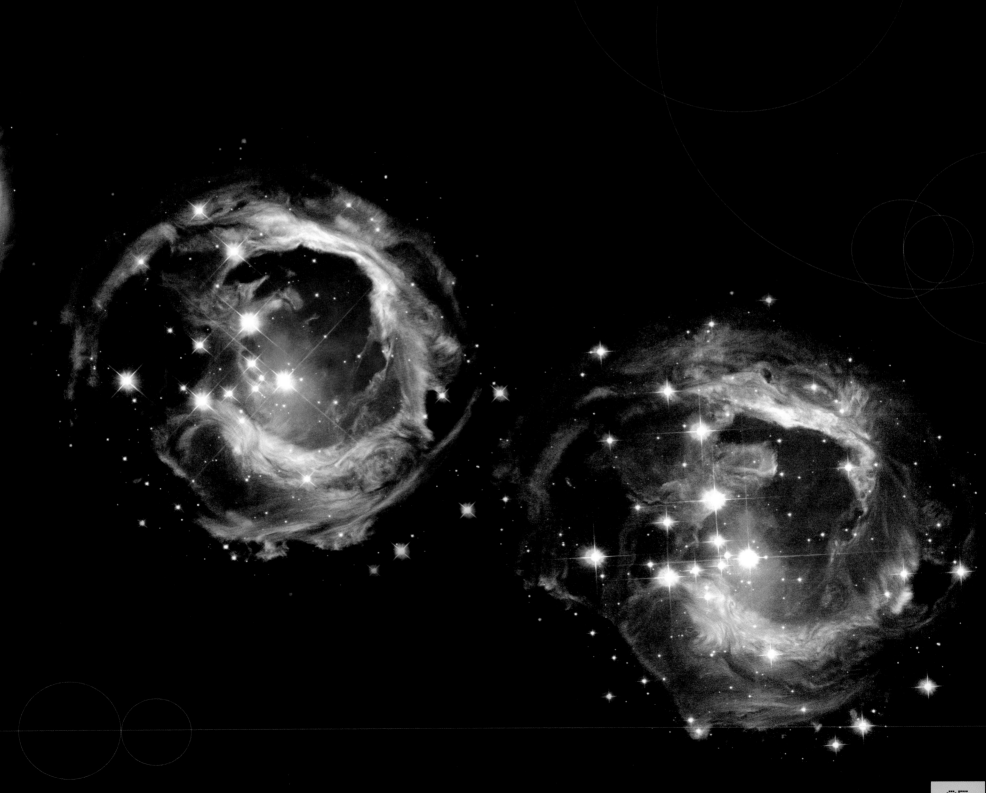

NAME: HD 47536

TYPE: RED GIANT
PLANETARY SYSTEM

DISTANCE: 395 LIGHT YEARS

NOTES: THIS RED GIANT HAS SWOLLEN
TO 23 TIMES THE SUN'S DIAMETER AND
SHINES WITH 100 TIMES OUR STAR'S
LUMINOSITY. A GIANT PLANET, FIVE
TIMES THE MASS OF JUPITER, ORBITS
THE STAR ONCE EVERY 712 DAYS.
0101001010110
DOWNLOAD COMPLETED
10101100110
CONTINUE
Y/N?

DATA PACKET ACCEPTED. 010001001101100010011100101001001001001001

Betelgeuse is one of the brightest red giants visible from Earth, and it is the only star large enough for special techniques to reveal features on its disc, including a distinctive "hot spot". The star is so large that it is unstable, fluctuating in size between 300 and 400 times the diameter of the Sun, with an average brilliance of 14,000 Suns. With roughly 15 solar masses of material, Betelgeuse is technically a supergiant that will destroy itself in a supernova explosion, but it offers a rare detailed view of a dying giant.

<< EXPAND SEARCH
> BETELGEUSE

86 672DARWY7001029

> ROTTEN EGG NEBULA

This distorted cloud of expanding gas is a dying red giant star caught in the act of developing into a planetary nebula with a central white dwarf. As the envelope of burning gas around the core swells in size, it causes the star's outer layers to expand violently, puffing away into space. Meanwhile, the star's collapsing core heats up, blowing out new layers of faster-moving gas that form a glowing supersonic shock wave as they escape from the star at 1 million kph

672DARWY76...KD77672DARWY73910-PPE>>8637

DYING SUNS

[INCOMING DATA... ACCEPTED]

WHEN A STAR EXHAUSTS THE FUEL IN ITS CORE, IT STARTS TO BURN THROUGH THE GAS IN ITS OUTER LAYERS, CAUSING IT TO BECOME UNSTABLE. SUCH A STAR MAY SHINE FAR MORE BRIGHTLY THAN BEFORE, BUT IT ALSO BALLOONS IN SIZE AND ITS SURFACE COOLS UNTIL IT BECOMES A RED GIANT.

NAME: HELIX NEBULA
DESIGNATION: NGC 7293

DISTANCE: 450 LIGHT YEARS

DIAMETER: 2.5 LIGHT YEARS

NOTES: THE CLOSEST PLANETARY
NEBULA TO EARTH. EACH OF THE GAS
GLOBULES ON THE INNER EDGE IS
ROUGHLY THE DIAMETER OF OUR
SOLAR SYSTEM.

01010010110
DOWNLOAD COMPLETED
10101100110
CONTINUE
Y/N?

« EXPAND SEARCH

› ANT NEBULA

The central star of the Ant Nebula is surrounded by a pair of white lobes, with a hurricane of high-speed gas flowing out around them at speeds of around 3.6 million kph. So-called "double-lobed" nebulae may be shaped by an unseen second star orbiting within the nebula, or they may be single stars with strong magnetic fields directing their gas.

DESIGNATION: MENZEL 3
LENGTH: 1.6 LIGHT YEARS
DISTANCE: 3,000 LIGHT YEARS

DATA PACKET ACCEPTED. 010010110110110

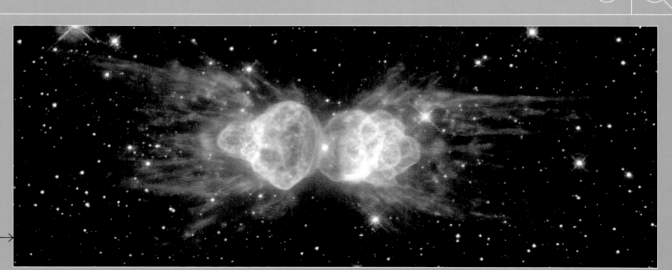

> ESKIMO NEBULA

The unusual Eskimo Nebula looks like a face peering out from within a furry hood. The roots of its "fur" are slow-moving gas globules that escaped from the central star about 10,000 years ago. Today, these globules are being stripped of material by faster-moving gas blowing out from the central star. The central region of the nebula is in fact a double-lobed planetary nebula seen almost end-on.

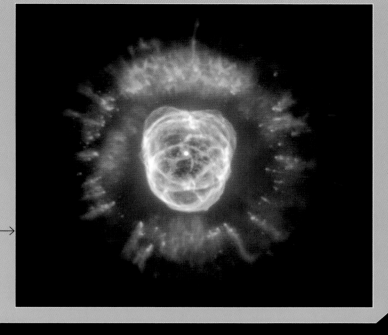

DESIGNATION: NGC 2392
DIAMETER: 0.7 LIGHT YEARS
DISTANCE: 5,000 LIGHT YEARS

01010010110
DOWNLOAD COMPLETED

> RED RECTANGLE

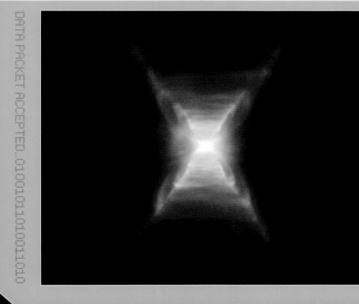

This bizarre planetary nebula is one of the strangest sights in the sky because the universe does not often create squares or rectangles. In fact, the nebula is a pair of conical jets that, by chance, we observe side-on. The jets are created by a pair of tightly bound stars at the centre of the nebula – gas thrown out from the dying star is funnelled into cones by a dense ring of material around the orbit of its companion.

DESIGNATION: HD 44179
LENGTH: 0.4 LIGHT YEARS
DISTANCE: 2,300 LIGHT YEARS

01010010110
DOWNLOAD COMPLETED

672DARWY7001...KD-PPE>>86371020

PLANETARY NEBULAE

INCOMING DATA... ACCEPTED >

PLANETARY NEBULAE ARE GHOSTLY STELLAR CAST-OFFS, QUITE DIFFERENT FROM STARBIRTH NEBULAE. THEY FORM WHEN AN UNSTABLE RED GIANT STAR OF ROUGHLY THE SUN'S MASS NEARS THE END OF ITS LIFE AND FLINGS OFF ITS OUTER LAYERS OF MATERIAL. OFTEN THEY FORM RINGS OR SPHERICAL SHELLS OF GLOWING GAS THAT RESEMBLE PLANETS (WHICH IS WHY THEY ARE CALLED PLANETARY NEBULAE), BUT EXTERNAL INFLUENCES CAN FUNNEL, TWIST AND BLOW THEM INTO BIZARRE AND BEAUTIFUL SHAPES.

NAME: CAT'S
EYE NEBULA
DESIGNATION: NGC 6543

DISTANCE: 3,300 LIGHT YEARS
LENGTH (INNER RING):
0.5 LIGHT YEARS
DIAMETER (OUTER HALO):
0.8 LIGHT YEARS

NOTES: DISCOVERED BY WILLIAM
HERSCHEL IN 1786. FIRST NEBULA TO
HAVE ITS SPECTRUM RECORDED.

01010010110
DOWNLOAD COMPLETED
10101100110
CONTINUE
Y/N?

≪ IMAGE ENHANCE

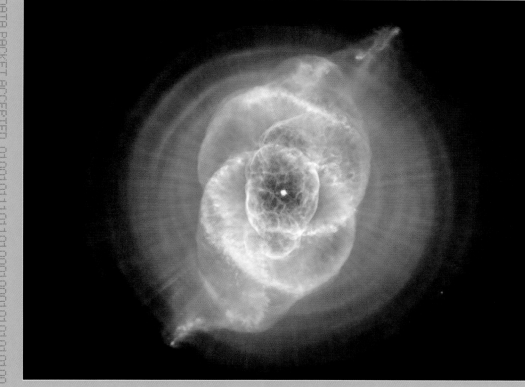

DATA PACKET ACCEPTED...0100101101101001000100101.01.00

This processed image of the Cat's Eye Nebula reveals the presence of concentric rings of gas beyond the inner nebula. These evenly spaced "shells" were ejected by the nebula's red-giant predecessor as it went through a series of unstable pulsations. The image also reveals that the central nebula is a series of overlapping bubbles. The presence of a hidden binary companion may pull the bubbles in different directions.

Strong stellar winds have "burst" the outer bubbles at each end.

CAT'S EYE NEBULA

INCOMING DATA... ACCEPTED >

THE CAT'S EYE NEBULA IS ONE OF THE MOST BEAUTIFUL PLANETARY NEBULAE IN THE SKY AND ALSO ONE OF THE MOST STUDIED. ITS COMPLEX STRUCTURE IS A RESULT OF INTERACTIONS BETWEEN MATERIAL RECENTLY EJECTED BY THE DYING STAR AND HOTTER BUT LESS DENSE GAS FLUNG OUT AT AN EARLIER STAGE. THE COMPLEX SPIRAL STRUCTURE OF THE NEBULA'S HEART SUGGESTS THE PRESENCE OF A CLOSE BINARY STAR AT THE CENTRE.

DATA PACKET ACCEPTED...0100101101101001001001.001010010100101010010101.00

An X-ray close-up of the nebula's central star reveals that it is surrounded by a cloud of extremely hot gas. A hot stellar wind, at several million degrees, blows out from the star's surface, causing the previously ejected gas bubbles to balloon outwards. The star itself is also hot enough to give off X-rays – it was the first such X-ray source to be discovered in a planetary nebula.

The strength of colour in this image shows the strength of the X-rays.

SUPERGIANTS

INCOMING DATA... ACCEPTED >

THE MOST MASSIVE STARS OF ALL ARE SUPERGIANTS. THEY BURN MANY THOUSANDS OF TIMES MORE BRIGHTLY THAN THE SUN, BUT THEY PAY FOR THEIR BRILLIANCE WITH VERY SHORT LIFE CYCLES THAT LAST JUST A FEW MILLION YEARS. AS THEY NEAR THE END OF THEIR NATURAL LIVES, THE STARS BECOME VIOLENT AND UNSTABLE. THEY SURROUND THEMSELVES WITH THICK CLOUDS OF MATERIAL FLUNG OFF FROM THEIR OUTER ATMOSPHERES, WHICH OFTEN MASK THE STARS THEMSELVES FROM SIGHT.

672DARWY7001...KD-PPE>>

DATA PACKET ACCEPTED..01.001.011.011.01.1.01.0001.001.01.01.001.01.0001.01.0110

The Eta Carinae supergiant is at the heart of the Carina Nebula, a star-forming region. It became the second brightest star in the sky during a brief eruption in the 1840s, before it dimmed – it is now no longer visible to the naked eye. The Homunculus Nebula, a double-lobed cloud of debris produced by the eruption, hides the star from view. Eta Carinae may, in fact, be a pair of supergiant stars, each 60 to 80 times the mass of the Sun.

DISTANCE:
8,000 LIGHT YEARS
LUMINOSITY:
1.5 MILLION SUNS (VISIBLE LIGHT),
5 MILLION SUNS OVERALL

01010010110
DOWNLOAD COMPLETED
10101100110
CONTINUE
Y/N?
10101100101
10101100101

> ETA CARINAE
≪ EXPAND SEARCH

NAME: BUBBLE NEBULA
CLASSIFICATION: NGC 7635

DISTANCE: 7,100 LIGHT YEARS

NOTES: THIS TURBULENT BUBBLE OF GAS IS BEING INFLATED BY STELLAR WINDS FROM A GIANT STAR 40 TIMES THE MASS OF THE SUN AND MORE THAN 400,000 TIMES AS LUMINOUS.

01010010110
DOWNLOAD COMPLETED
10101100110
CONTINUE
Y/N?

> PISTOL STAR

One of the most luminous stars in our entire galaxy, the Pistol Star pumps out as much energy in 20 seconds as the Sun does in a year. It is a blue "hypergiant", roughly the diameter of Earth's orbit, and its surface is so hot that it cannot hold itself together, boiling off into space and creating the nebula that now surrounds it. Despite its brilliance, the Pistol Star is only visible in infrared (heat) radiation because it is hidden behind clouds of gas and stars near the centre of the galaxy.

DATA PACKET ACCEPTED..0100100101011010101010000010101110

DISTANCE: 25,000 LIGHT YEARS
LUMINOSITY: 1.7 MILLION SUNS
OVERALL

01010010110
DOWNLOAD COMPLETED
10101100110
CONTINUE Y/N?
10101100101 10101100101

NAME: VEIL NEBULA
DESIGNATION: NGC 6960

DISTANCE: 2,600 LIGHT YEARS

NOTES: THE VEIL NEBULA IS THE
BRIGHTEST PART OF THE "CYGNUS
LOOP", THE REMAINS OF A SUPERNOVA
THAT EXPLODED 15,000 YEARS AGO.
THIS GLOWING SHOCK WAVE FORMED AS
THE REMNANT'S EXPANDING HOT GAS
PLOUGHED INTO COOLER GAS AROUND IT.

01010010110
DOWNLOAD COMPLETED
10101100110
CONTINUE
Y/N?

SUPERNOVAE AND THEIR REMNANTS

INCOMING DATA... ACCEPTED >

STARS WITH AT LEAST EIGHT TIMES THE MASS OF THE
SUN END THEIR LIVES IN ENORMOUS EXPLOSIONS
CALLED SUPERNOVAE. AS THEY BURN THROUGH ALL
THEIR REMAINING FUEL IN JUST A FEW DAYS OR WEEKS,
SUPERNOVAE CAN BRIEFLY OUTSHINE ENTIRE GALAXIES
BEFORE SUBSIDING TO LEAVE AN EXPANDING CLOUD OF
HOT, GLOWING GAS CALLED A REMNANT.

<< EXPAND SEARCH
> SN 1987A

DATA PACKET ACCEPTED..01001110

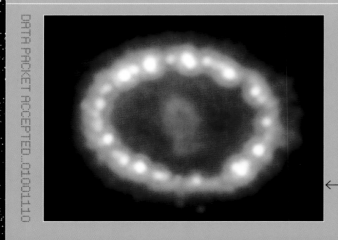

Supernovae are rare events – there has not been one seen
in our Milky Way Galaxy for about four centuries. The closest
and brightest recent supernova erupted in 1987 in the
Large Magellanic Cloud, one of our neighbouring galaxies.
Discovered on 23 February, it soon became the brightest
star in its entire galaxy, taking several months to fade away.
Astronomers eventually linked its origin to the explosion of
a blue supergiant star.

*This Hubble image shows knots of glowing
gas squeezed and excited by the supernova's
expanding shock wave.*

DISTANCE: 6,300 LIGHT YEARS
TOTAL ENERGY OUTPUT: 750,000 SUNS

DOWNLOAD COMPLETED

CONTINUE
Y/N?

The Crab Nebula is the glowing remnant of a supernova whose light reached Earth in 1054 and was recorded by Chinese, Arab and Native American stargazers. The cloud of shredded gas has since grown to 10 light years across, still illuminated from within by radiation from the Crab Pulsar (*see* page 96).

CRAB NEBULA
« DATA FILE

« EXPAND SEARCH
> CASSIOPEIA A

DATA PACKET ACCEPTED.010010010010101011001000

This supernova remnant is extremely faint in visible light but is one of the strongest sources of radio waves in the sky. It is the remnant of an explosion that occurred about 350 years ago, but for some reason (perhaps dark dust clouds blocked the view), it was not recorded by astronomers on Earth. Cassiopeia A also gives off powerful X-rays from its outer edge, where temperatures can reach as high as 30 million°C.

DISTANCE: 10,000 LIGHT YEARS
COMPOSITION: OXYGEN,
SULPHUR, SILICON AND IRON

01010010110
DOWNLOAD COMPLETED

> G11.2-0.3

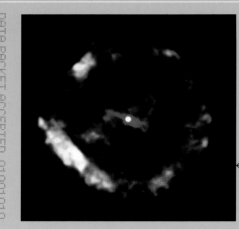

Most neutron stars are also pulsars. These are stars whose radiation appears to flash on and off hundreds of times a second. Pulsars form because the magnetic field of a collapsed neutron star is powerful enough to channel the star's escaping radiation into narrow beams that sweep across the sky as the star rotates. G11.2-0.3 is one of the slowest known pulsars, flashing at just 14 times a second.

DISTANCE: 16,000 LIGHT YEARS
SUPERNOVA OBSERVED IN AD386
01010010110
DOWNLOAD COMPLETED
10101100110

→ Combined X-ray and visible light images reveal where the stellar wind interacts with the magnetic field.

The pulsar at the heart of the Crab Nebula spins 30 times a second, pumping out beams of light from its magnetic poles, while fierce stellar winds blow material off its equator. Particles in the wind give off radiation of their own as they interact with the star's powerful magnetic field, illuminating the surrounding nebula.

> CRAB PULSAR

<< DATA FILE

This planetary nebula only recently formed when a dying red giant star puffed off its outer layers to reveal its still-hot core, with a surface temperature of 200,000°C thanks to residual heat. As white dwarfs cool over many millions of years, they shrink and fade and eventually become black dwarfs.

> ## NGC 2440
≪ DATA FILE

WHITE DWARFS, NEUTRON STARS AND BLACK HOLES

INCOMING DATA... ACCEPTED >

WHETHER A STAR DIES IN A GENTLE PLANETARY NEBULA OR A VIOLENT SUPERNOVA, IT ALWAYS LEAVES BEHIND ITS HOT DENSE CORE. DEPENDING ON ITS MASS AND THE FORCE OF ITS COLLAPSE, THE CORE MAY FORM EITHER AN EARTH-SIZED WHITE DWARF, A CITY-SIZED NEUTRON STAR OR A BLACK HOLE – A REGION OF SUPERDENSE MATTER FROM WHICH NOT EVEN LIGHT CAN ESCAPE.

DATA PACKET ACCEPTED...01001011101010001010110101010100010001011010

Sirius B is the white dwarf companion of Sirius A, which is the brightest star in Earth's sky. Sirius B has roughly the mass of the Sun contained in a volume slightly smaller than Earth's. The star that formed Sirius B must once have shone even brighter than Sirius A itself.

Sirius B still contains half the mass of Sirius A, but shines just one-10,000th as brightly.

> ## SIRIUS B
≪ EXPAND SEARCH

NAME: THE MILKY WAY
GALAXY

DIAMETER: 100,000 LIGHT YEARS

NOTES: THIS ARTIST'S IMPRESSION
DEPICTS THE VIEW FROM ABOVE THE
MILKY WAY ON THE SUN'S SIDE OF THE
GALAXY. THE DISC AT THE CENTRE AND
THE SPIRAL ARMS ARE CLEARLY SEEN.

01010010110
DOWNLOAD COMPLETED
10101100110
CONTINUE
Y/N?

<< DESTINATION
> THE SOLAR SYSTEM

<< IMAGE ENHANCE
> PLANE OF THE MILKY WAY

DATA PACKET ACCEPTED .0100010

The Sun and its solar system lie roughly two-thirds of the way from the centre of the galaxy between two spiral arms called Orion and Perseus. The galaxy is rotating, and it takes roughly 230 million years for the Sun to orbit the galactic centre. When we look out into space, we are looking across the plane of the galaxy, so the spiral arms appear in the sky as a narrow, winding band of pale star clouds.

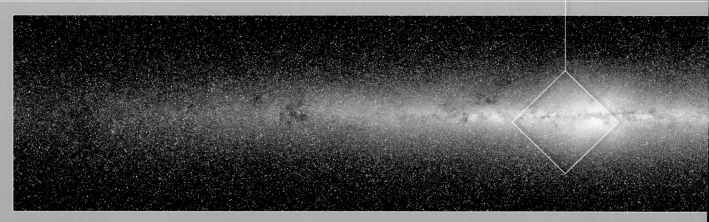

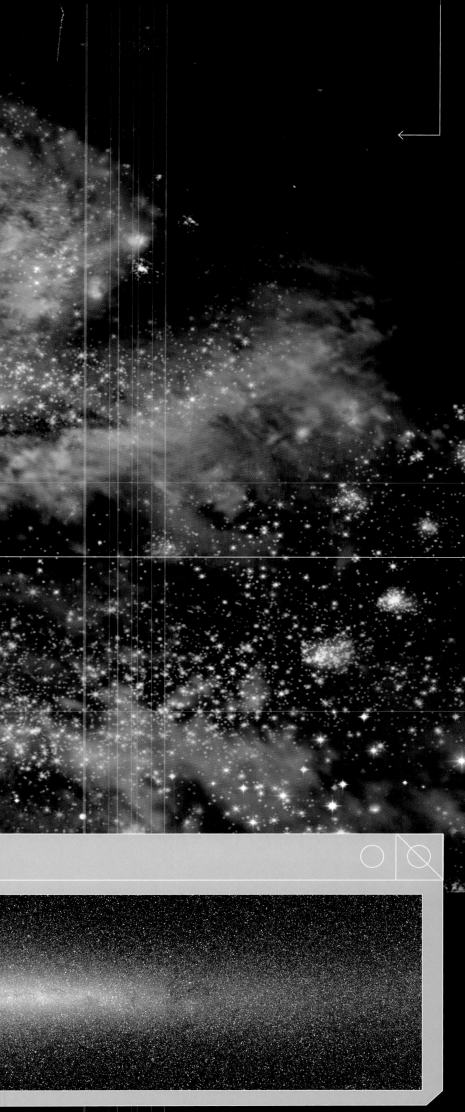

OUR GALAXY

OUR SOLAR SYSTEM AND ALL THE STARS IN THE SKY ARE PART OF A SINGLE MIGHTY STAR SYSTEM, OR GALAXY, CALLED THE MILKY WAY. OUR GALAXY IS A SLOWLY SPINNING SPIRAL CONTAINING ABOUT 200 BILLION STARS, AS WELL AS HUGE AMOUNTS OF GAS, DUST AND OTHER MATTER.

The Milky Way's spiral arms emerge from a disc of fainter stars. Along the leading edge of each arm, there are dark dust clouds and bright nebulae, both signs of star formation in action. Behind them are the bright, short-lived young stars of countless open star clusters (groups of stars). These silhouette the dust clouds and define the shape of the spiral arms themselves. Long-lived stars, such as the Sun, may not shine as brightly, but they survive long enough for their orbits to take them out of the spiral arms and into the disc.

Long-exposure photographs show dark dust lanes silhouetted against densely packed star clouds.

> STARS, GAS AND DUST
« IMAGE ENHANCE

NAME: GALACTIC
CENTRE

DISTANCE: 26,000 LIGHT YEARS

NOTES: THIS INFRARED IMAGE
PIERCES THE VEIL OF DUST AND STARS
TO REVEAL THE GALACTIC CENTRE. OLD
COOL STARS APPEAR BLUE, WHILE DUST
HEATED BY HOT YOUNG STARS IS RED. THE
VERY HEART OF THE MILKY WAY APPEARS
AS A WHITE GLOW.

0101011110
DOWNLOAD COMPLETED
1010110110
CONTINUE
Y/N?

> RADIATION SIGNATURE ANALYSIS

« RADIATION ANALYSIS

This X-ray image reveals a large number of bright objects in the central region of the Milky Way. Some are black holes, others are brilliant, supermassive stars and others are hot stellar remnants, such as white dwarfs. The enveloping mist is caused by sparse but hot gas heated to millions of degrees by its violent surroundings. The exact centre, or core, is hidden behind the bright central glare – it corresponds to a source of radio waves known as Sagittarius A*.

DATA PACKET ACCEPTED.0110100101000110001100

CENTRE OF THE MILKY WAY

INCOMING DATA... ACCEPTED >

THE HEART OF THE MILKY WAY LIES 26,000 LIGHT YEARS FROM EARTH, HIDDEN BEHIND DENSE STAR CLOUDS IN THE CONSTELLATION SAGITTARIUS. THESE CLOUDS CONCEAL AN ENORMOUS BLACK HOLE, WITH THE MASS OF MILLIONS OF SUNS, WHOSE INFLUENCE RULES THE CENTRAL REGIONS.

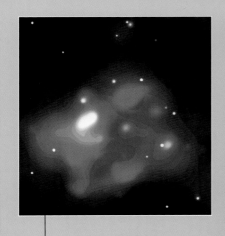

These X-ray-emitting clouds form part of an X-ray "ridge" running above the galactic centre.

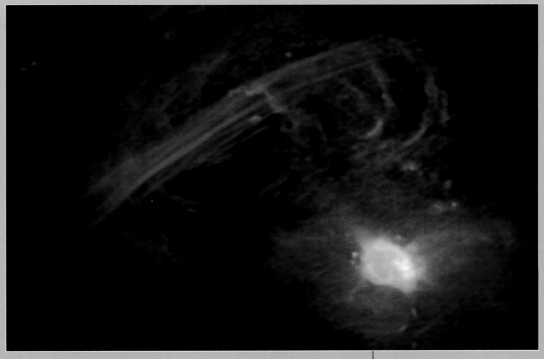

Individual filaments in the radio arc are a few light years wide but 100 light years long.

The region around the central black hole is filled with bizarre high-energy objects. The galactic centre radio arc (right) traces streams of particles emerging from close to Sagittarius A* itself (opposite) and interacting with the galaxy's own magnetic field. X-ray-emitting clouds form where the streams of particles in the radio arc slam into cooler gas near the galaxy's core and release energy (left). Elsewhere, densely packed clusters of giant stars surround the supermassive black hole. However, this black hole itself gives off only a gentle radio signal – it reveals its presence more clearly through its effect on the orbits of nearby stars.

> DETAILS IN THE CORE
« RADIATION ANALYSIS

DATA PACKET ACCEPTED.011010101010001110101010101001100011101

BEYOND OUR GALAXY

THROUGH THE GALAXIES

Our galaxy is just one among many – by some estimates, there are as many galaxies in the Universe as there are stars in the Milky Way. Some of these galaxies resemble our own – they are huge, slowly rotating spirals of stars. Others are equally large, or even larger, balls known as ellipticals. But most are comparatively small clumps of just a few million or a few billion stars – dwarf galaxies that are either elliptical or irregular in shape.

Different types of galaxy have different compositions – spirals have a mix of ancient yellow and red stars in their hub, bright blue stars in their spiral arms, and gas, dust and middle-aged stars like the Sun in the disc between the arms. Ellipticals are composed mostly of long-lived red and yellow stars, and have little gas and dust to form new generations of short-lived blue and white stars. Irregulars, in contrast, have huge amounts of gas to create new generations of stars, and so are dominated by blue and white stars.

The composition of a galaxy also helps determine its shape. Compared to their size, individual stars are usually separated by huge regions of space, so collisions and close encounters between them are rare – and these events are what is needed to flatten a galaxy into a rotating disc. As a result, galaxies that have little gas in them have stars in a variety of chaotic orbits, forming an overall elliptical shape. In spiral galaxies, dust and gas clouds collide with each other far more often, and their random motions soon even out into roughly circular orbits in a flattened disc. Since the gas clouds in turn give rise to the stars, it's no surprise that the stars also form a disc.

Each type of galaxy represents a stage in evolution – part of a story we are only just working out. Irregular galaxies seem to merge together into enormous clouds that then develop spiral structures as a natural result of the way they rotate. Spirals grow by absorbing smaller galaxies, but occasional major collisions between them can be enough to disrupt their structure, throwing their stars into the haphazard orbits of an elliptical. While the merged galaxies still retain enough gas in orbit around them, they can form an intermediate kind of galaxy known as a lenticular, before eventually restarting star formation and regenerating their spiral arms. However, repeated collisions can strip away their star-forming gas, heating it up so that it moves fast enough to escape their gravity altogether. With no more gas to form new stars, the galaxy will remain an elliptical forever more.

Galaxies of all sizes often form around huge black holes at their centres. These "supermassive" black holes, sometimes with the weight of a million or more Suns, are far too big to have formed in stellar explosions – they probably began as collapsing gas clouds, growing larger by pulling in any material that strayed too close. Active galaxies, which often have unusually bright cores or emit jets of high-speed particles, are created by black holes that are still feeding on their surroundings.

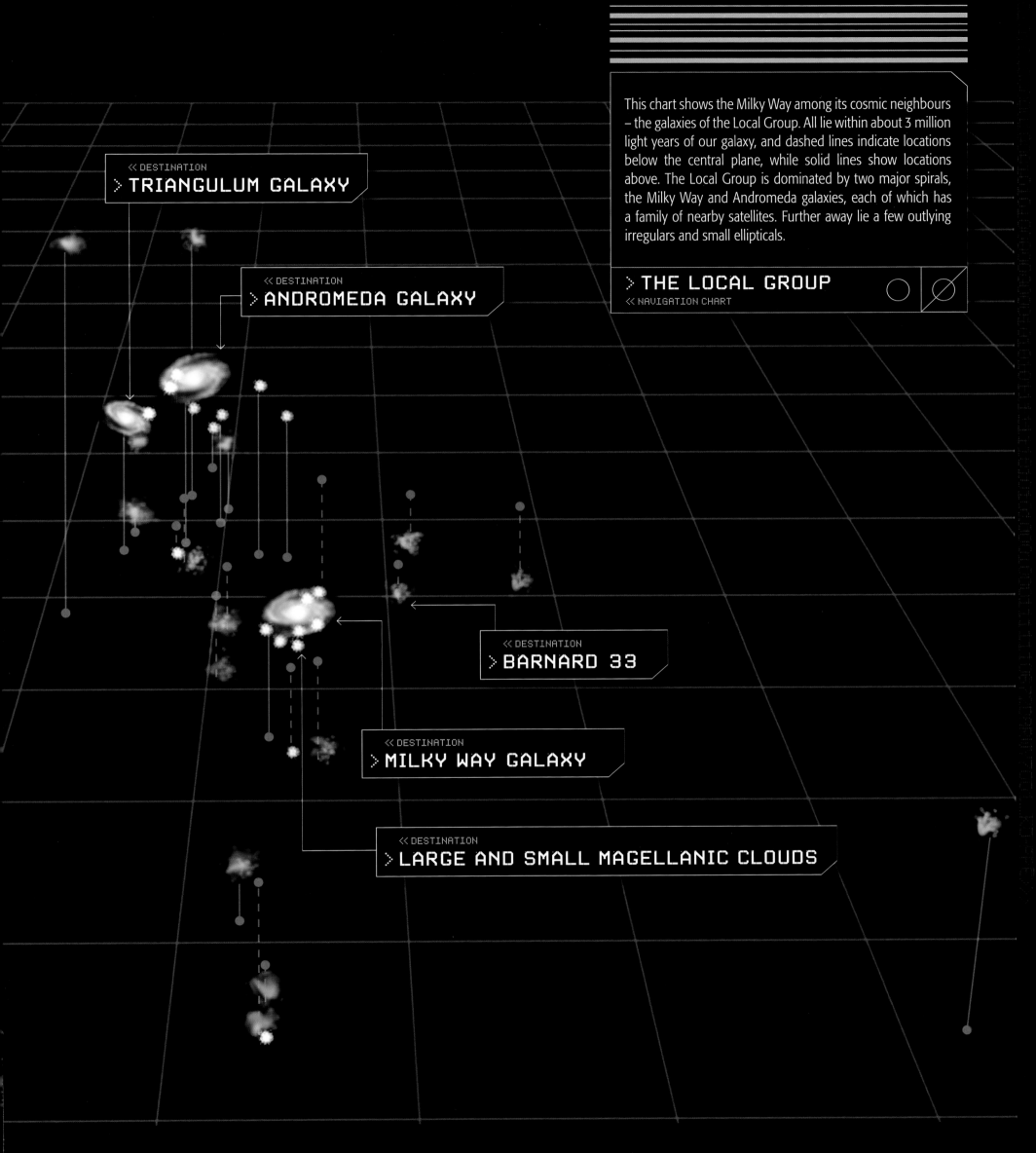

This chart shows the Milky Way among its cosmic neighbours – the galaxies of the Local Group. All lie within about 3 million light years of our galaxy, and dashed lines indicate locations below the central plane, while solid lines show locations above. The Local Group is dominated by two major spirals, the Milky Way and Andromeda galaxies, each of which has a family of nearby satellites. Further away lie a few outlying irregulars and small ellipticals.

> THE LOCAL GROUP
<< NAVIGATION CHART

<< DESTINATION
> BARNARD 33

<< DESTINATION
> MILKY WAY GALAXY

<< DESTINATION
> LARGE AND SMALL MAGELLANIC CLOUDS

NAME:
TARANTULA NEBULA
CLASSIFICATION: NGC 2070

DIAMETER: 1,000 LIGHT YEARS
DISTANCE: 180,000 LIGHT YEARS

NOTES: THE TARANTULA NEBULA IN THE
LMC IS SO LARGE AND BRIGHT THAT, IF
IT WAS AS CLOSE TO EARTH AS THE ORION
NEBULA, IT WOULD BE BRIGHT ENOUGH TO
CAST SHADOWS AND COVER 1/12TH OF
THE ENTIRE SKY

01010010110
DOWNLOAD COMPLETED
10101100110
CONTINUE
Y/N?

LARGE MAGELLANIC CLOUD

INCOMING DATA... ACCEPTED >

THE LARGE MAGELLANIC CLOUD (LMC)
IS THE BRIGHTEST OF THE GALAXIES
CLOSE TO THE MILKY WAY. IT IS A
FAIRLY SHAPELESS CLOUD OF STARS,
GAS AND DUST, ALTHOUGH TRACES OF
A BAR OF STARS ACROSS ITS CORE
SUGGEST THAT IT MAY HAVE ONCE
HAD A SINGLE SPIRAL ARM. THE
LMC AND ITS SMALLER COMPANION,
THE SMALL MAGELLANIC CLOUD
(SMC), ARRIVED IN OUR GALACTIC
NEIGHBOURHOOD IN ABOUT THE LAST
BILLION YEARS.

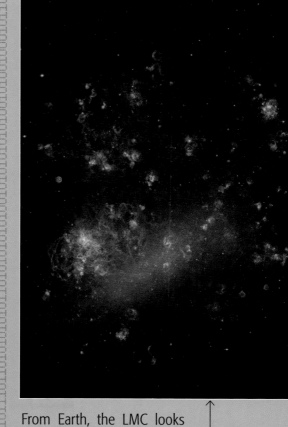

DATA PACKET ACCEPTED. 01001011101101011010101010100110010010010110010010010110

From Earth, the LMC looks
like a fragment of the Milky
Way that has broken off.
In fact, the Large and Small
Magellanic Clouds are not
true satellites of our own
galaxy. They are probably
making their first close
approach to the Milky Way
from intergalactic space.

*The LMC
contains more
than 10 billion
stars, including
the wreckage
of Supernova
1987A.*

> WIDE VIEW
<< EXPAND SEARCH

Shock waves from the cluster's supernovae disturb nearby clouds of gas.

This star cluster in the LMC is called Hodge 301 and is on the outskirts of the enormous Tarantula Nebula. It is dominated by brilliant blue-white stars that are doomed to short lifespans. Their presence shows that the entire cluster is very young, perhaps just 25 million years old. Since its formation, it has already drifted 150 light years from the centre of the nebula, where the next generation of stars has begun to form a new cluster. At least 40 high-mass stars in Hodge 301 have already exploded into supernovae.

DATA PACKET ACCEPTED.010010010010101010101010010

<< IMAGE ENHANCE

> HODGE 301

NAME: SMALL
MAGELLANIC CLOUD
DESIGNATION: SMC

DIAMETER: 10,000 LIGHT YEARS
DISTANCE: 200,000 LIGHT YEARS

NOTES: KNOTS OF BRIGHT GAS FORM THE
BRIGHTEST AREAS OF THE SMC. MANY OF
THE BRIGHTEST STARS IN THIS IMAGE
ACTUALLY LIE IN THE FOREGROUND,
WITHIN THE MILKY WAY.

01010010110
DOWNLOAD COMPLETED
10101100110
CONTINUE
Y/N?

SMALL MAGELLANIC CLOUD

INCOMING DATA... ACCEPTED >

THE SMALL MAGELLANIC CLOUD (SMC) IS BOTH SMALLER AND SLIGHTLY MORE DISTANT THAN THE LMC, BUT IT IS EQUALLY RICH IN STAR-FORMING REGIONS. BOTH GALAXIES ARE FOLLOWED BY A TRAIL OF GAS AND STARS CALLED THE MAGELLANIC STREAM. THIS CONSISTS OF MATERIAL THAT IS BEING TORN AWAY FROM THE GALAXIES AS THEY FALL UNDER THE INFLUENCE OF THE MILKY WAY'S POWERFUL GRAVITY FIELD.

<< IMAGE ENHANCE
> N90 AND NGC 602

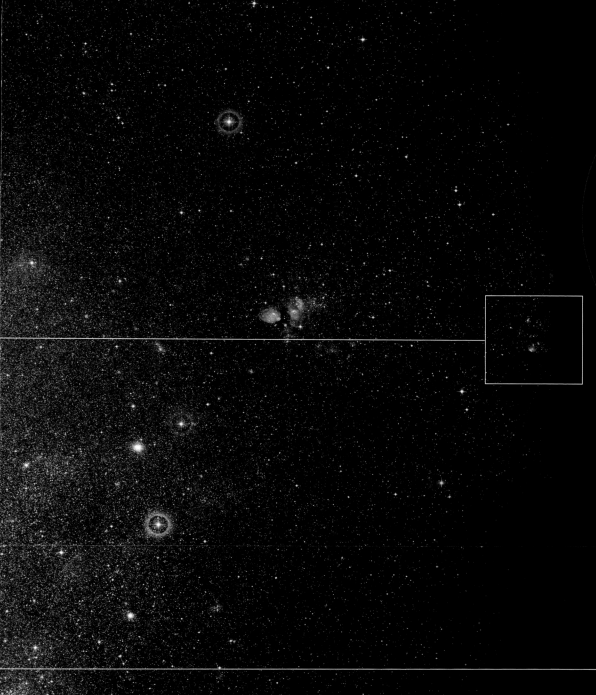

DATA PACKET ACCEPTED..01001011101010101010101010

The oldest stars in NGC 602 began to shine about 5 million years ago.

A small but brilliant star cluster, NGC 602, illuminates a glowing "cavern" of gas known as N90. This seems to be the SMC's equivalent of the Milky Way's Eagle Nebula, complete with towering pillars and hanging "stalactites" of gas and dust, which are being eroded by the fierce radiation of the young stars. Star formation here is still continuing – as stellar winds from the centre have pushed out through the nebula, star formation has gradually moved out towards the edges.

DATA PACKET ACCEPTED..01001011101010001011101000110

More than half of the SMC's most massive stars are concentrated in the NGC 346 cluster at the heart of a star-forming region known as N66. There are three distinct "subclusters" containing dozens of massive stars. The young stars are still trapped in the surrounding gas, which is sculpted into sheets and swirls by the hot stellar winds. The dense outer areas of the N66 nebula prevent the gas escaping.

Hydrogen clouds (top) and young stars (bottom).

DIAMETER: 400 LIGHT YEARS
DISTANCE: 200,000 LIGHT YEARS
PHOTOGRAPHED BY THE HUBBLE SPACE

> NGC 346 AND N66
« IMAGE ENHANCE

NAME: ANDROMEDA
GALAXY
DESIGNATION: M31

DIAMETER: 140,000 LIGHT YEARS
DISTANCE: 2.5 MILLION LIGHT YEARS

NOTES: LIKE THE MILKY WAY, ANDROMEDA
HAS ITS OWN SYSTEM OF SATELLITE
GALAXIES. THE MOST OBVIOUS OF THESE
ARE THE SMALL ELLIPTICAL GALAXIES
M32 AND M110.

01010010110
DOWNLOAD COMPLETED
10101100110
CONTINUE
Y/N?

DATA PACKET ACCEPTED. 010010010010101011011010

This infrared image reveals cool dust clouds in Andromeda's complex rings.

This X-ray image shows Andromeda's supermassive black hole, which probably has the mass of 30 million Suns.

Astronomers have learned a lot about Andromeda by studying different wavelengths of radiation. Infrared images of the galaxy reveal a bright, lopsided ring of star formation about halfway from the core. They also show that Andromeda contains about twice as many stars as the Milky Way. X-ray images show a dozen or more bright yellow X-ray sources around the core. Each source is likely to be a black hole in its own right. Andromeda's supermassive central black hole (the blue source near the centre) is surprisingly cool.

>> RADIATION SIGNATURE ANALYSIS

ANDROMEDA SPIRAL

INCOMING DATA... ACCEPTED >

THE ANDROMEDA GALAXY IS THE NEAREST MAJOR GALAXY TO OUR OWN. IT IS AN ENORMOUS SPIRAL EVEN LARGER THAN THE MILKY WAY AND THE MOST DISTANT OBJECT THAT CAN BE SEEN FROM EARTH WITH THE NAKED EYE. ALTHOUGH ANDROMEDA IS ABOUT 2.5 MILLION LIGHT YEARS AWAY, GRAVITY IS PULLING IT AND THE MILKY WAY TOGETHER AT ABOUT 500,000 KPH. IN ABOUT 3 BILLION YEARS, THE GALAXIES WILL COLLIDE AND EVENTUALLY MERGE.

DATA PACKET ACCEPTED. 01001011011010001010010110101001101001

Image enhancements reveal two distinct concentrations of stars at the core of the Andromeda Galaxy. The fainter of the two dense "nuclei" lies at the true centre of Andromeda, while the brighter one is 5 light years off-centre. Andromeda may contain not one but two supermassive black holes at its heart, although the division may be an illusion created by a dust cloud.

Each "nucleus" is a dense ball containing millions of old red and yellow stars.

> **ANDROMEDA'S CORE**
<< IMAGE ENHANCE

NAME: TRIANGULUM
GALAXY
DESIGNATION: M33

DIAMETER: 60,000 LIGHT YEARS
DISTANCE: 2.9 MILLION LIGHT YEARS

NOTES: THE TRIANGULUM GALAXY
HAS THE SAME MASS AS 40 BILLION
SUNS, SO IS FAR LIGHTER THAN ITS
NEIGHBOURING SPIRAL GALAXIES.

01/10010110

DOWNLOAD COMPLETED
10101100110
CONTINUE
Y/N?

<< IMAGE ENHANCE

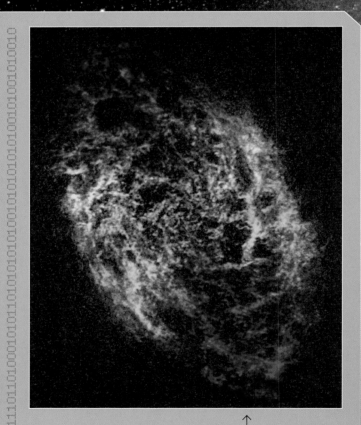

DATA PACKET ACCEPTED .01001011101010001010101010100110010010010010

As a galaxy rotates, the wavelength and colour of the light from its stars change. Thanks to a phenomenon called the Doppler effect, an object moving away from Earth appears redder and one getting closer to Earth looks bluer. The faster an object moves in relation to Earth, the greater the "Doppler shift" in its colour.

The colours in this image of the Triangulum show that different parts of the galaxy spin at different rates.

> DOPPLER SHIFT
<< IMAGE ENHANCE

9RUY7001029UDGRT56...K07767ZDARUY7391D-PPE>>6637102010284720139UYHF7463921

THE TRIANGULUM GALAXY

INCOMING DATA... ACCEPTED >

THE TRIANGULUM GALAXY IS THE THIRD MAJOR MEMBER OF THE CLUSTER OF GALAXIES CALLED THE LOCAL GROUP, BUT IT IS ONLY ABOUT HALF THE SIZE OF THE OTHER TWO – THE MILKY WAY AND ANDROMEDA. IT IS CLOSE TO ANDROMEDA AND MAY BE IN ORBIT AROUND IT. TRIANGULUM IS FAINTER AND MORE CLUMPY THAN SOME OTHER SPIRALS. IT IS A "FLOCCULENT" GALAXY IN WHICH THE SPIRAL ARMS ARE WEAKENED AND DISRUPTED, PERHAPS BY ANDROMEDA'S GRAVITY.

DATA PACKET ACCEPTED..01000101011011010

Triangulum is home to one of the largest starbirth nebulae known – NGC 604. It lies in one of the galaxy's outlying spiral arms and is larger than any nebula in the Milky Way, at about 1,500 light years across. The nebula's structure is shaped by radiation from more than 200 hot bright stars, each with the mass of 15 Suns or more. Their fierce ultraviolet light causes the surrounding gas to glow brightly.

NGC 604 is more than 6,000 times brighter than the Orion Nebula in the Milky Way.

‹‹ IMAGE ENHANCE

> NGC 604

NAME:
PINWHEEL GALAXY
DESIGNATION: M101

DIAMETER: 170,000 LIGHT YEARS
DISTANCE: 27 MILLION LIGHT YEARS

NOTES:
THIS BEAUTIFUL FACE-ON SPIRAL
IS ONE OF THE LARGEST SPIRAL
GALAXIES KNOWN. ITS OPEN ARMS
REVEAL COUNTLESS BRIGHT STAR
CLUSTERS AND DARK DUST LANES.

01010010110
DOWNLOAD COMPLETED
10101100110
CONTINUE
Y/N?

SPIRAL GALAXIES

INCOMING DATA... ACCEPTED >

ABOUT A QUARTER OF ALL NEARBY GALAXIES ARE SPIRALS. THESE SLOWLY ROTATING COSMIC PINWHEELS HAVE ARMS DEFINED BY BRIGHT KNOTS OF BLUE-WHITE STARS AND A BALL-SHAPED HUB OF OLDER YELLOW AND RED STARS. THE SPIRAL ARMS ARE NOT FIXED, BUT ARE CONTINUOUSLY REGENERATED AS NEW STARS FORM IN A ROTATING SPIRAL PATTERN AND THE OLDER STARS EXPLODE OR FADE AWAY.

« EXPAND SEARCH
> NGC 4414

DATA PACKET ACCEPTED. 010010111011001101100

This beautiful galaxy is an example of a flocculent spiral, similar to the Triangulum Galaxy. When the normal processes that trigger spiral waves of star formation in a galaxy are cut off, stars instead form in clumps, triggered by local events such as collisions between nebulae or shock waves from supernovae. In NGC 4414's case, these clumps have spread across the galaxy's disc.

Dark lanes of dust trace the galaxy's spiral pattern in silhouette.

« EXPAND SEARCH
> SOMBRERO GALAXY

DATA PACKET ACCEPTED. 010010111011010011010

This spiral galaxy, roughly 50 million light years from Earth in the constellation Virgo, lies almost edge-on from our point of view. Blazing stars silhouette a narrow dust lane that runs all the way around the galaxy, forming a distinct rim. This, combined with the large, bulging hub around the supermassive black hole at the centre, makes the galaxy look a little like a sombrero, a traditional Mexican hat.

A dense swarm of more than 2,000 globular clusters orbit above and below the Sombrero's disc.

NAME: NGC 1300
AND IC 4703

DIAMETER: 100,000 LIGHT
YEARS OR MORE
DISTANCE: 69 MILLION
LIGHT YEARS

NOTES: THIS LARGE BARRED SPIRAL IS
S-SHAPED AND HAS AN UNUSUAL MINI-
SPIRAL, 3,300 LIGHT YEARS ACROSS,
AT ITS CENTRE.

DOWNLOAD COMPLETED

CONTINUE
Y/N?

BARRED SPIRALS

INCOMING DATA... ACCEPTED >

MANY SPIRAL GALAXIES CONTAIN A BAR OF STARS AND GAS. THIS EXTENDS IN BOTH DIRECTIONS FROM THE CENTRAL HUB, WITH THE SPIRAL ARMS EMERGING FROM ITS ENDS. THE BAR IS NO MORE SOLID THAN THE SPIRAL ARMS, BUT IS JUST AN EFFECT CAUSED BY STARS FOLLOWING ELONGATED ELLIPTICAL ORBITS RATHER THAN PERFECT CIRCLES. ASTRONOMERS HAVE RECENTLY DISCOVERED THAT OUR OWN MILKY WAY GALAXY IS A BARRED SPIRAL.

672DARWY7001...KD-PPE >>

DIAMETER: 82,000 LIGHT YEARS
DISTANCE: 183 MILLION LIGHT YEARS

01010010110
DOWNLOAD COMPLETED
10101100110

This barred spiral in the constellation Pavo has a normal structure in visible light but ultraviolet light reveals hidden features. This picture combines a visible light image with an ultraviolet one (in blue). Ultraviolet is given off by objects too hot to emit much visible light – in this image, it is emitted by a ring of star formation around the galaxy's core linked to the central bar at two points. Farther out, loose patches of bright, hot stars trace the galaxy's two spiral arms.

« EXPAND SEARCH
> NGC 6782

This colour-coded image of the centre of the barred spiral NGC 1512 in the Horologium constellation reveals detailed features of the galaxy's interior. In the main image, the colours emphasize the natural properties of stars – hot, bright stars are blue, cooler fainter stars are red. Though invisible in this image due to its faintness, the bar crosses the central ring. It is thought to funnel gas into a circular region of star formation, highlighted by the presence of short-lived but brilliant stars.

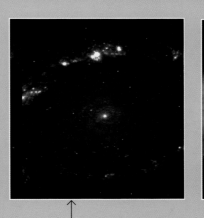

This ultraviolet image reveals hot stars in the ring and centre of the hub.

This infrared image shows dust and cool stars around the hub.

« EXPAND SEARCH
> RADIATION ANALYSIS NGC 1512

KD77672DARWY73910-PPE >> 6637102010234720I9 117

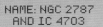

NAME: NGC 2787
AND IC 4703

CORE DIAMETER:
4,500 LIGHT YEARS
DISTANCE: 24 MILLION LIGHT YEARS

NOTES: THIS BARRED LENTICULAR
GALAXY HAS A CORE SURROUNDED BY
TIGHTLY WOUND DUST LANES, BUT NO
SIGNS OF SPIRAL ARMS.

01010010110
DOWNLOAD COMPLETED
10101100110
CONTINUE
Y/N?

M60 is a giant elliptical close to the centre of the Virgo Cluster of galaxies. This enhanced image shows it next to a more distant and probably unrelated spiral galaxy. Large ellipticals are almost always found near the centre of major galaxy clusters – the only regions where galaxy collisions and mergers are frequent enough to allow them to grow.

DIAMETER:
120,000 LIGHT
YEARS
DISTANCE:
60 MILLION
LIGHT YEARS
LUMINOSITY:
EQUIVALENT
TO 60 BILLION
SUNS.
01010010110
DOWNLOAD
COMPLETED
10101100110
CONTINUE
Y/N?

> M60
« EXPAND SEARCH

DATA PACKET ACCEPTED. 0101011010100011101010101010101011010

ELLIPTICAL AND LENTICULAR GALAXIES

INCOMING DATA... ACCEPTED >

ELLIPTICAL GALAXIES ARE THE MOST COMMON TYPE AND ARE FOUND
IN A GREATER RANGE OF SIZES THAN SPIRALS. THEY INCLUDE GIANT
ELLIPTICALS AND DWARF GALAXIES. ALL ARE DOMINATED BY YELLOW AND
RED STARS AND LACK STAR-FORMING GAS AND DUST. LENTICULAR GALAXIES
HAVE AN ELLIPTICAL HUB SURROUNDED BY A DISC OF GAS BUT NO SPIRAL
ARMS OR OBVIOUS NEW STAR FORMATION.

> DIAMETER: 215,000 LIGHT YEARS
  DISTANCE: 75 MILLION LIGHT YEARS
  01010010110
  DOWNLOAD COMPLETED
  10101100110
  CONTINUE
  Y/N?

Also known as Fornax A, this lenticular galaxy shrouded in thin veils of dust is a strong source of radio waves. The dust is probably the remains of one or more spirals recently swallowed up by the larger galaxy. The cosmic collision has transformed NGC 1316 into an active galaxy (*see* page 126).

> NGC 1316
<< DATA FILE

<< EXPAND SEARCH
> M87

This giant elliptical galaxy has more than a trillion stars. Such galaxies contain very little star-forming gas and are dominated by long-lived red and yellow stars. M87 lies at the heart of the Virgo Cluster of galaxies and probably formed and grew by swallowing up and merging with other galaxies. Gas and dust from recently consumed galaxies remain around its core, feeding the central supermassive black hole and making M87 an active galaxy.

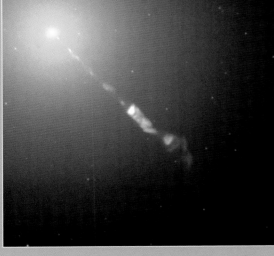

DATA PACKET ACCEPTED.01001010011010

As M87's black hole feeds on gas and dust, it spits out a jet of high-speed particles.

Small irregular galaxies have weaker gravity than larger star systems, so they can find it hard to hold onto hot, fast-moving gas. NGC 4214 in the constellation Canes Venatici is a good example. While most of its stars are comparatively old and stable, it is studded with bright clusters of star formation, where the powerful stellar winds from hot young stars have blown the nearby gas into filaments. The most massive stars in these clusters have already destroyed themselves in supernovae, creating enormous bubbles in the surrounding nebulae.

DATA PACKET ACCEPTED..01001001001011011010

NAME: CIGAR GALAXY
DESIGNATION: M82

DIAMETER: 40,000 LIGHT YEARS
DISTANCE: 12 MILLION LIGHT YEARS

NOTES: AN ENORMOUS WAVE OF STAR FORMATION LIGHTS UP THIS STARBURST GALAXY FOLLOWING ITS CLOSE ENCOUNTER WITH A NEIGHBOUR 600 MILLION YEARS AGO.

01010010110
DOWNLOAD COMPLETED
10101100110
CONTINUE
Y/N?

DATA PACKET ACCEPTED...010010111011010001011010110101001010010

Also known as NGC 6822, Barnard's Galaxy is a bright irregular member of the Local Group. Like the Large Magellanic Cloud, it has traces of a central bar and countless star-forming nebulae. The presence of the bar shows that this galaxy is just large enough to start developing a more complex structure.

DIAMETER:
10,000 LIGHT YEARS
DISTANCE: 1.8 MILLION LIGHT YEARS
01010010110
DOWNLOAD COMPLETED
10101100110
CONTINUE Y/N?
10101100101
10101100101

> **BARNARD'S GALAXY**
« EXPAND SEARCH

IRREGULAR GALAXIES

INCOMING DATA... ACCEPTED >

AS THEIR NAME SUGGESTS, IRREGULAR GALAXIES HAVE LITTLE OR NO OBVIOUS STRUCTURE. TYPICALLY A FEW THOUSAND LIGHT YEARS ACROSS, THEY ARE USUALLY VERY RICH IN GAS, DUST AND BRIGHT YOUNG STARS. OFTEN, THEY HAVE ENORMOUS WAVES OF STAR FORMATION KNOWN AS STARBURSTS. IRREGULAR GALAXIES ARE THOUGHT TO BE COSMIC BUILDING BLOCKS THAT GROW INTO LARGER, MORE STRUCTURED SPIRAL GALAXIES BY MERGING AND COLLIDING WITH EACH OTHER.

NAME: WHIRLPOOL
GALAXY
DESIGNATION: M51

DIAMETER: 38,000 LIGHT YEARS
DISTANCE: 23 MILLION LIGHT YEARS

NOTES: M51'S CLOSE COMPANION NGC
5195 CRASHED THROUGH THE LARGER
GALAXY ABOUT HALF A BILLION YEARS
AGO AND NOW LIES JUST BEHIND IT. THE
ENCOUNTER HELPED REINFORCE M51'S
WELL-DEFINED SPIRAL ARMS.

01010010110
DOWNLOAD COMPLETED
10101100110
CONTINUE
Y/N?

≪ EXPAND SEARCH

> SEYFERT'S SEXTET

DATA PACKET ACCEPTED: 01001010001101110

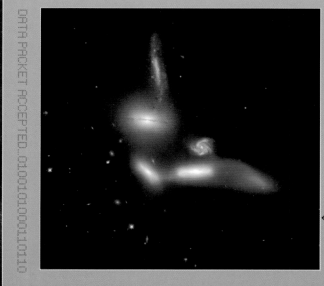

This closely packed galaxy group, 190 million light years away, actually consists of four galaxies not six. The face-on spiral is really five times farther away than the other galaxies and what appears to be a galaxy on the far right is really a tail torn away from one of the other galaxies by tidal forces. The four true galaxies are closely packed into just 100,000 light years, dancing around each other before eventually merging.

The distorted shapes of the galaxies show they are already beginning to tear apart.

0101001011010101101011010101001011010101100101010011010011001011000101010100101010101101011010101001011010011010101001011010011001010100101001101001100101100010

DATA PACKET ACCEPTED..01001011011001011000...

DIAMETER: 100,000 LIGHT YEARS
DISTANCE: 150 MILLION LIGHT YEARS
01010010110
DOWNLOAD COMPLETED
10101100110
CONTINUE
Y/N?
10101100101 10101100101

This edge-on spiral galaxy in the constellation Hydra has a buckled appearance due to a recent close encounter with another galaxy. Eventually, the galaxy's proper shape will return. The encounter has not only pulled the galaxy's disc out of shape, but has also triggered new star formation where tidal forces have pulled on orbiting gas clouds and forced them to compress. The youngest clusters are highlighted by clouds of short-lived blue stars.

> **ESO 510-G13**
≪ EXPAND SEARCH

INTERACTING GALAXIES

INCOMING DATA... ACCEPTED >
CLOSE ENCOUNTERS BETWEEN GALAXIES ARE COMMON ACROSS THE UNIVERSE BECAUSE, FOR THEIR SIZE, THEY ARE QUITE CLOSELY PACKED TOGETHER. WHEN GALAXIES PASS EACH OTHER, THEIR POWERFUL GRAVITY CAN RAISE TIDES THAT TRIGGER NEW WAVES OF STAR FORMATION AND MAY AFFECT THEIR STRUCTURE. SOME GALAXIES ARE PULLED INTO ORBIT AROUND EACH OTHER, EVENTUALLY COLLIDING AND MERGING TO FORM GIANT ELLIPTICAL GALAXIES

DESIGNATION:
AM 0644-741

DIAMETER: 150,000 LIGHT
YEARS
DISTANCE: 300 MILLION LIGHT
YEARS

NOTES: THIS RING GALAXY HAS SUFFERED
A SIMILAR COLLISION TO THE CARTWHEEL
GALAXY. AS THE SMALLER GALAXY PASSED
THROUGH THE LARGER ONE, THE SPIRAL
ARMS WERE DISRUPTED BY A SPREADING
RING OF STAR FORMATION.

01010010110
DOWNLOAD COMPLETED
10101100110

COLLIDING GALAXIES

INCOMING DATA... ACCEPTED >

INTERGALACTIC COLLISIONS ARE PART OF THE COSMIC
LIFE CYCLE. THEY ALLOW GALAXIES TO GROW AND
TRANSFORM THEM FROM ONE TYPE TO ANOTHER. WHEN
GALAXIES COLLIDE, MOST OF THE STARS PASS BY EACH
OTHER UNSCATHED, BUT THE CLOUDS OF GAS AND DUST
ARE COMPRESSED AND HEATED BY POWERFUL FORCES TO
IGNITE ENORMOUS FIRES OF STAR FORMATION.

67 2455DMRW7001...KD+PPE>>>

> CARTWHEEL GALAXY

« EXPAND SEARCH

DATA PACKET ACCEPTED 00101110110

This colourful image reveals a galaxy with a core and an outer ring of bright stars linked together by faint lines of material that look like the spokes of a wheel – giving the galaxy its name. The Cartwheel Galaxy is suffering the aftermath of a hit-and-run collision with a smaller galaxy (one of the three at the lower left) 100 million years ago. The culprit was travelling fast enough to escape from the scene of the crime.

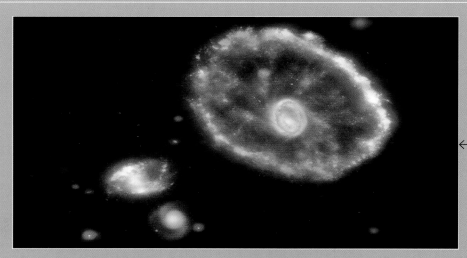

This picture combines ultraviolet (blue), X-ray (purple), visible light (green) and infrared (red) images to reveal the form of the Cartwheel Galaxy.

DIAMETER (TOTAL): 400,000 LIGHT YEARS
DIAMETER (CORE): 60,000 LIGHT YEARS
DISTANCE: 63 MILLION LIGHT YEARS

The colliding spirals NGC 4038 and 4039 are known as the Antennae because their spiral arms have unwound into long streamers that resemble an insect's antennae. Meanwhile, gas clouds rammed together in the heart of the collision have erupted in an enormous blaze of star formation between the galaxies' cores.

> ANTENNAE GALAXIES
≪ DATA FILE

> BLACK EYE GALAXY
≪ EXPAND SEARCH

DATA PACKET ACCEPTED…01.001.01.001.001.001.0011.00111.10

This dust-laden spiral, 50,000 light years across, is still recovering from its last meal – a substantial galaxy that it swallowed whole roughly a billion years ago. All that remains of the smaller galaxy is an outer halo of gas rotating in the opposite direction to the rest of the galaxy.

Where the invisible halo rubs against the Black Eye's outer edge, it triggers new starbirth within the dark dust clouds.

NAME: NGC
3079
TYPE: SEYFERT GALAXY

DIAMETER: 70,000 LIGHT YEARS
DISTANCE: 50 MILLION LIGHT
YEARS

NOTABLE FEATURES:
THIS SEYFERT GALAXY HAS AN
UNUSUALLY BRIGHT CORE AND IS
ALSO GOING THROUGH A RAPID BURST
OF STAR FORMATION, PRODUCING
FIERCE RADIATION. THIS CAUSES
BUBBLES OF GAS, SOME 3,000
LIGHT YEARS ACROSS, TO
BLOW OUT FROM AROUND
ITS CENTRE.

ACTIVE GALAXIES

MANY GALAXIES HAVE ENERGY SOURCES AT
THEIR CENTRES THAT SHINE TOO BRIGHTLY
AND CHANGE TOO RAPIDLY TO BE CAUSED BY STARS
ALONE. THESE ARE CALLED ACTIVE GALAXIES, AND
THEY RANGE FROM QUIET, NEARBY SEYFERT GALAXIES
TO BRILLIANT BUT DISTANT QUASARS AND BLAZARS. EACH
TYPE OF ACTIVE GALAXY HAS THE SAME POWER SOURCE AT
ITS HEART: A BLACK HOLE FEEDING ON GAS, DUST AND THE
SHREDDED REMAINS OF STARS.

<< EXPAND SEARCH
> ## HE0450-2958

This brilliant "star" is, in fact, a distant quasi-
stelllar object or quasar – the most violent
type of active galaxy. A quasar's blazing light
comes from the disc of superheated material
being dragged into its enormous central black
hole. Usually, a quasar is surrounded by a
large but faint "host galaxy", but in the case of
HE0450-2958, the quasar seems to be alone
in space – perhaps pushed out of its orginal
host galaxy by a galactic collision.

DISTANCE: 3 BILLION LIGHT YEARS
MASS: AT LEAST 400 MILLION SUNS

01010010110
DOWNLOAD COMPLETED
10101100110
CONTINUE Y/N?
10101100101 10101100101

DATA PACKET ACCEPTED...0100100101010101010000

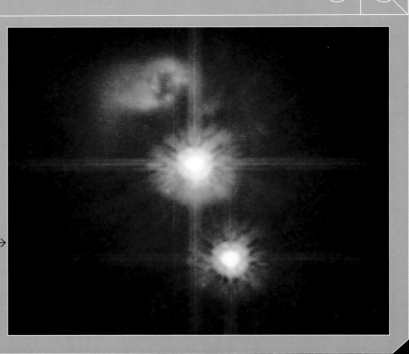

DATA PACKET ACCEPTED.0100100101010110101000001011100001011100010

Cygnus A gives off one of the brightest radio signals in the entire sky from its two lobes of glowing gas. The power source of this "radio galaxy" lies hidden at the centre of the picture (right), in the heart of a dim elliptical galaxy. From there, jets of high-speed particles shoot out across 300,000 light years of space, billowing into a pair of bright "lobes" as they encounter nearby invisible gas clouds and warm them until they glow at radio wavelengths.

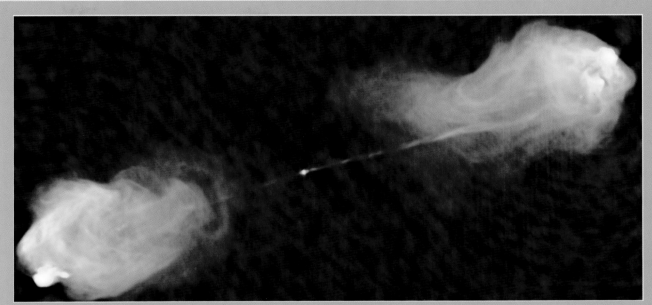

DESIGNATION: 3C 405.0
DIAMETER: 600,000 LIGHT YEARS
DISTANCE: 700 MILLLION LIGHT YEARS

01010010110
DOWNLOAD COMPLETED
10101100110
CONTINUE Y/N?

X-ray images of Cygnus A (left) reveal more details of its structure. The entire system lies within a larger galaxy cluster and is surrounded by a region of hot gas that is giving off X-rays. The galaxy's twin jets are creating an expanding cavity within this gas that is shaped rather like an American football or a rugby ball. The points where the jets balloon out into the galaxy's radio lobes are marked by X-ray "hot spots" on either side.

DISC DIAMETER: 800 LIGHT YEARS
DISTANCE: 100 MILLION LIGHT YEARS

DOWNLOAD COMPLETED

CONTINUE
Y/N?

This active elliptical galaxy reveals a secret that most other active galaxies keep hidden – the central black hole is surrounded by a doughnut-shaped ring of dust that can block out the brilliant radiation from around the hole itself. Different types of active galaxy are the result of us seeing this central region at different angles.

› NGC 4261
≪ DATA FILE

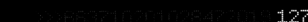

NAME: CENTAURUS A
DESIGNATION: NGC 5128

DIAMETER: 60,000 LIGHT YEARS
DISTANCE: 15 MILLION LIGHT YEARS

NOTABLE FEATURES:
AS THE MAIN GALAXY ABSORBS THE
REMAINS OF ITS VICTIM, COMPRESSED
GAS CLOUDS GENERATE BRILLIANT
BLUE-WHITE STAR CLUSTERS ALONG
THE EDGES OF THE DARKER DUST LANE.

01010010110
DOWNLOAD COMPLETED
10101100110
CONTINUE
Y/N?

CENTAURUS A

INCOMING DATA... ACCEPTED >

THE ELLIPTICAL GALAXY NGC 5128 IS ONE OF THE BRIGHTEST AND NEAREST ACTIVE GALAXIES, OFTEN KNOWN BY THE NAME OF ITS RADIO SOURCE, CENTAURUS A. THE DARK DUST LANE ACROSS THE MIDDLE OF THE GALAXY SHOWS WHERE IT HAS RECENTLY SWALLOWED A SMALLER SPIRAL GALAXY, AND ASTRONOMERS THINK IT WAS THIS COLLISION THAT SPARKED CENTAURUS A'S CURRENT WAVE OF ACTIVITY.

An infrared image of Centaurus A's centre reveals the complex hidden structure of the central dust lane as a four-sided shape called a parallelogram. This appearance is a result of the spiral galaxy's flat disc being twisted out of shape by the powerful gravity of the larger elliptical galaxy. The merger probably began about 200 million years ago, and as it continues, the orbiting dust cloud will flatten out once again before it is absorbed completely.

Multiwavelength image of Centaurus A

Infrared image of Centaurus A from the Spitzer Space Telescope.

DATA PACKET ACCEPTED. 01001001010101011101010

<< EXPAND SEARCH

> ## SPITZER IMAGE

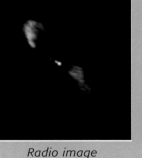

X-ray image *Radio image* *Visible light image*

Photographs of high-energy radiation reveal the extent of Centaurus A's activity as well as traces of an enormous explosion in the galaxy's core, probably about 10 million years ago. Two enormous arcs of hot gas emerge from the central region, the most prominent parts of a ring 25,000 light years across. X-rays also reveal jets of high-energy particles still emerging from the core, while radio waves show where the jets form lobes as they plough into surrounding gas clouds.

DATA PACKET ACCEPTED. 0100100101010101010110101010

<< EXPAND SEARCH

> ## HIGH-ENERGY OUTBURST

NAME: NGC 3314

DIAMETER: c.70,000 LIGHT
YEARS EACH
DISTANCE: 117 AND 140 MILLION
LIGHT YEARS

NOTES: THE STRANGE APPEARANCE OF
NGC 3314 HAS A SIMPLE EXPLANATION
– IT IS REALLY TWO GALAXIES THAT
HAPPEN TO LIE IN THE SAME DIRECTION,
SO THAT THE OUTLINE OF ONE IS
SILHOUETTED AGAINST THE OTHER.
01010010110
DOWNLOAD COMPLETED
10101100110
CONTINUE
Y/N?

« EXPAND SEARCH
> **NGC 4650A**

"Polar ring" galaxies such as NGC 4650A are like
spirals tipped on their sides. They have a central
core of long-lived red and yellow stars encircled
by a ring of younger blue and white stars. They
are probably the result of an unusual galaxy
collision, but another interesting feature is that the
core's gravity seems too weak to hold onto the
fast-moving stars of the outer ring. It seems likely
that the polar rings contain large amounts of "dark
matter" (*see* page 136) that hold them together.

DESIGNATION: NGC 4650A
RING DIAMETER: 60,000 LIGHT YEARS
DISTANCE: 130 MILLION LIGHT YEARS

01010010110
DOWNLOAD COMPLETED
10101100101 10101100101

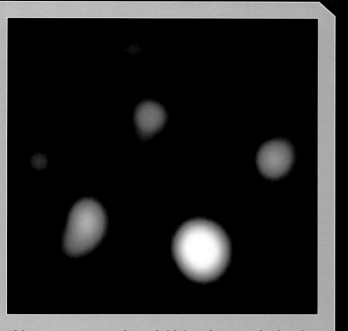

This strange cross-shaped blob, photographed using X-rays, is the result of an unseen foreground galaxy distorting the light from a more distant active galaxy to produce multiple images (*see* page 137). The effect is called an "Einstein cross" because it is caused by an effect of Albert Einstein's theory of relativity.

> **CLOVERLEAF QUASAR**
≪ EXPAND SEARCH

STRANGE GALAXIES

NOT ALL GALAXIES FIT NEATLY INTO THE CLASSIFICATION OF SPIRALS, ELLIPTICALS, LENTICULARS AND IRREGULARS. SOME HAVE DEVELOPED UNIQUE AND BEAUTIFUL SHAPES, USUALLY AS THE RESULT OF A COLLISION OR CLOSE ENCOUNTER WITH ANOTHER GALAXY. OTHER GALAXIES DON'T SEEM TO OBEY THE RULES AT ALL AND HAVE A STRANGE STRUCTURE OR COMPOSITION FOR REASONS THAT ASTRONOMERS CAN ONLY GUESS AT.

→ DESIGNATION: PGC 54559
RING DIAMETER: 100,000 LIGHT YEARS
DISTANCE: 600 MILLION LIGHT YEARS

DOWNLOAD COMPLETED

CONTINUE
Y/N?

Unlike "polar rings" and "cartwheels", true ring galaxies, such as Hoag's Object, have perfectly spherical cores of yellow stars, surrounded by bright and even rings of star formation. Astronomers still aren't sure how they form, but it probably involves one galaxy capturing material from another.

> **HOAG'S OBJECT**
≪ DATA FILE

DEPTHS OF THE UNIVERSE

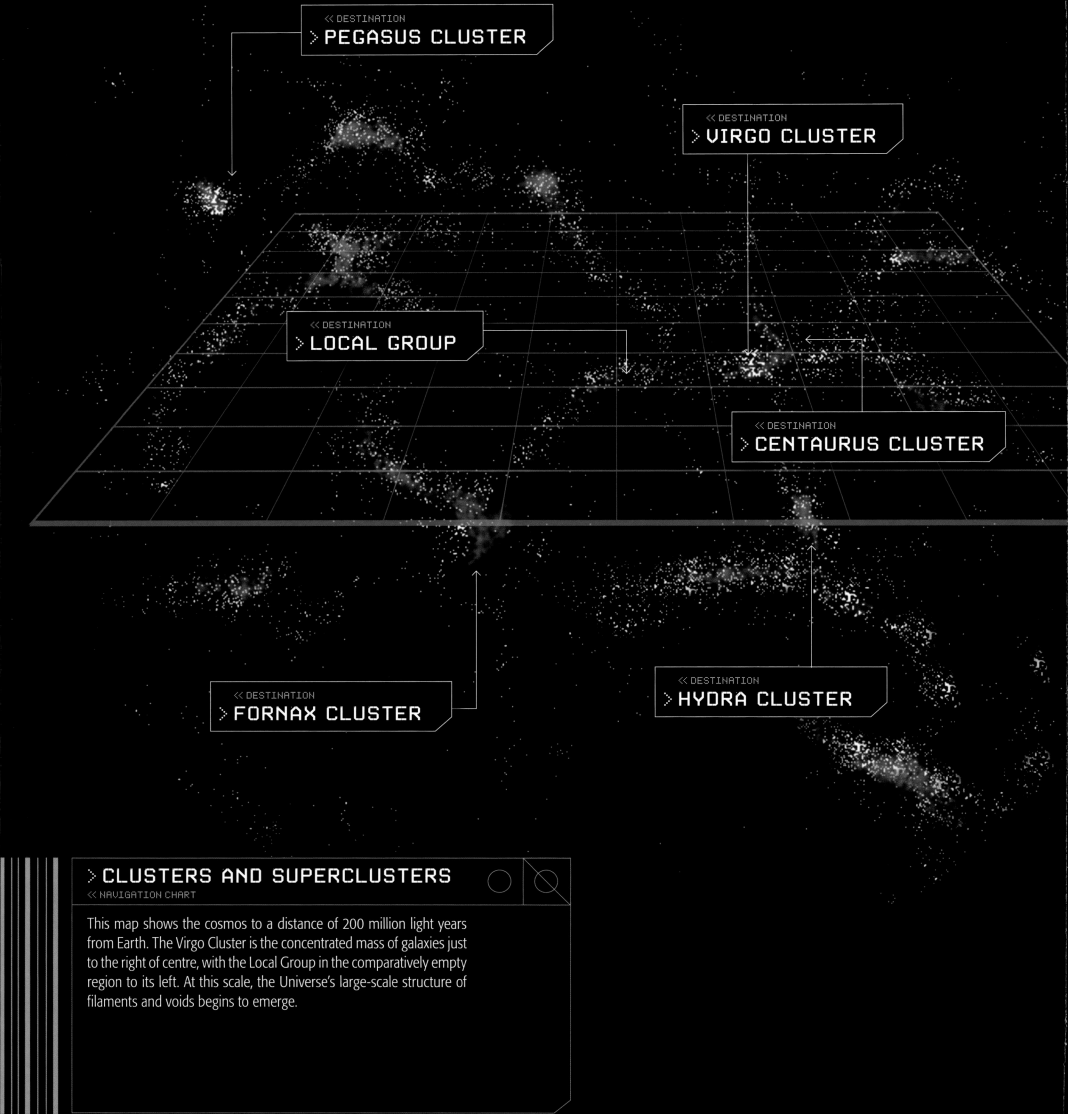

<< DESTINATION
> PEGASUS CLUSTER

<< DESTINATION
> VIRGO CLUSTER

<< DESTINATION
> LOCAL GROUP

<< DESTINATION
> CENTAURUS CLUSTER

<< DESTINATION
> FORNAX CLUSTER

<< DESTINATION
> HYDRA CLUSTER

> CLUSTERS AND SUPERCLUSTERS
<< NAVIGATION CHART

This map shows the cosmos to a distance of 200 million light years from Earth. The Virgo Cluster is the concentrated mass of galaxies just to the right of centre, with the Local Group in the comparatively empty region to its left. At this scale, the Universe's large-scale structure of filaments and voids begins to emerge.

ACROSS THE UNIVERSE

At the largest scale, the Universe is made up of countless galaxy clusters like our own Local Group. These clusters may contain anything from a few dozen to a few thousand small galaxies, held together by their own gravity. Despite the huge variety in the number of galaxies they contain, almost all clusters are about the same size – 5 or 10 million light years across.

Clusters merge and overlap at their edges, sometimes making it hard to tell which galaxies belong to which cluster. Together, they form superclusters, each centred on an extremely dense central cluster. Our small, lightweight Local Group lies on the outskirts of the Virgo Supercluster, which is centred on the heavyweight Virgo Cluster, about 60 million light years away.

And even superclusters blur into each other, creating enormous galaxy chains called filaments. Hundreds of millions of light years long, filaments are the largest structures in the Universe – they wrap themselves around equally huge and apparently empty holes called voids.

As we look out across the enormous gulf of space to observe distant galaxies, we are seeing light that may have taken millions or even billions of years to reach us. But when we measure and record the light from these galaxies, it reveals something else. The light is "red shifted" by the Doppler Effect (*see* page 112), implying that all these distant parts of the Universe are moving away from us. On its own, this might mean that Earth has a special place at the centre of everything, but something else is happening – the further away we look, the faster galaxies seem to be moving away. This reveals the true explanation: the entire Universe is expanding, carrying our galaxy along with it – the expansion just appears to be centred on us because of our point of view.

The fact that the Universe is expanding means that in the past, things were much closer together – in fact, when we "wind back" the expansion, it turns out that 13.7 billion years ago, everything would have been in exactly the same place, jammed together and heated to tremendous temperatures. This is the root of the Big Bang theory, our best explanation for the origin of the cosmos.

According to this theory, the Universe began in an enormous explosion 13.7 billion years ago – an event that spontaneously created all the matter and energy in the cosmos, as well as space and time themselves. As the Universe expanded rapidly, it cooled and became less dense, and swarms of tiny subatomic particles were able to join together to build atoms.

The early Universe was opaque – so densely packed with matter that light just bounced around in it like a fog. Eventually, 300,000 years after the Big Bang itself, the Universe became transparent. The light from this time can still be detected as a weak radio signal from every direction in the sky – it has travelled more than 13 billion years to reach Earth, and the opaque wall that it forms is an echo from the surface of the Universe when it was young.

NAME: DARK MATTER
COMPUTER SIMULATION

NOTES: THIS "MAP" OF DARK
MATTER DISTRIBUTION WAS CREATED
BY SIMULATING THE EVOLUTION
OF THE UNIVERSE AND MAKING
PREDICTIONS ABOUT HOW DARK MATTER
BEHAVES. THE PATTERNS CLOSELY MATCH
THE DISTRIBUTION OF VISIBLE MATTER
RECORDED BY 2MASS, SUGGESTING THAT
DARK MATTER PULLS OTHER MATERIAL
INTO SHAPE AROUND IT.

010100101110

DOWNLOAD COMPLETED
10101100110

CONTINUE
Y/N?

DATA PACKET ACCEPTED.01001001010101010101101

<< EXPAND SEARCH

> 2MASS SURVEY

Mapping the sky using infrared radiation reveals the structure of the Universe. This chart, produced by the Two Micron All Sky Survey (2MASS), plots the position of 1.6 million galaxies according to their location in the sky. The different colours indicate the distance of the galaxies from Earth. The galaxies form web-like "filaments" around large empty spaces called "voids" – a structure that is affected by the presence of dark matter.

Blue points are closest to Earth; red ones are furthest away.

DARK MATTER

INCOMING DATA... ACCEPTED >

ASTRONOMERS THINK THAT THERE SHOULD BE
A LOT MORE MATERIAL IN THE UNIVERSE THAN
WE CAN SEE WITH OUR TELESCOPES. THIS
"MISSING MASS" COULD ACCOUNT FOR UP TO 90
PER CENT OF ALL MATTER IN THE UNIVERSE.
ASTRONOMERS CALL IT "DARK MATTER", AND
IT COULD CONSIST OF DIM, UNDETECTABLE
OBJECTS SUCH AS ROGUE PLANETS, DEAD STARS
AND EVEN BLACK HOLES THAT LIE UNSEEN
WITHIN GALAXIES. IT COULD ALSO BE MADE
UP OF SUBATOMIC PARTICLES THAT CONTAIN
SIGNIFICANT AMOUNTS OF MASS, YET ARE
UNDETECTED BY OUR INSTRUMENTS.

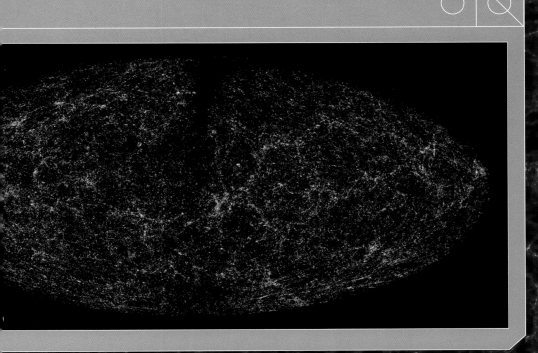

《 EXPAND SEARCH

⟩ GRAVITATIONAL LENSING

DATA PACKET ACCEPTED: 010010010110010010101010101010101010101010101010101.01

Concentrations of dark matter can give away their presence through their effect on the appearance of other objects. This is called "gravitational lensing" and happens when light rays from a distant object are deflected from their straight-line paths through space by passing close to a large, dense mass. The matter distorts space around it and bends the path of the light like a lens so that a distorted image reaches Earth. In this image, matter in the central galaxy cluster Abell 1689 is lensing the light from a more distant cluster, making it appear as arcs of light around the edges. The strength of such lensing effects shows that galaxy clusters have more mass than we would expect from the visible material in their galaxies. Part of this is in the form of hot gas clouds within the clusters, but most is unseen dark matter.

NUMBER OF GALAXIES: 3,000
DISTANCE: 2.2 BILLLION LIGHT YEARS
01010010110
DOWNLOAD COMPLETED
10101100110
CONTINUE Y/N?

THE DEEP UNIVERSE

INCOMING DATA... ACCEPTED >

BECAUSE EVEN LIGHT, THE FASTEST THING IN CREATION, TRAVELS AT
A LIMITED SPEED, THE UNIVERSE IS LIKE A GIANT TIME MACHINE. THE
FURTHER WE LOOK ACROSS SPACE, THE LONGER THE LIGHT WE SEE HAS
TAKEN TO REACH US AND THE FURTHER BACK IN TIME WE ARE OBSERVING.
NORMALLY, THIS DOES NOT MAKE A DIFFERENCE BECAUSE STARS AND
GALAXIES CHANGE SO SLOWLY THAT IT DOES NOT MATTER IF WE SEE
THEM AS THEY ARE NOW OR AS THEY WERE HUNDREDS OR EVEN MILLIONS
OF YEARS AGO. BUT TODAY'S LARGE TELESCOPES CAN DETECT TRACES
OF LIGHT ACROSS BILLIONS OF LIGHT YEARS, AND SO WE CAN LOOK AT
THE UNIVERSE WHEN IT WAS MUCH YOUNGER. PEERING BACK INTO THE
YOUNG COSMOS, WE CAN SEE THE FIRST GENERATIONS OF STARS AND THE
FORMATION OF GALAXIES IN THE AFTERMATH OF THE BIG BANG.

Looking across the depths of space has another strange effect – because the universe is still expanding from the burst of the Big Bang (*see* page 140), the more distant an object is, the faster it is moving away from us. And, just as the speed of a moving police siren affects the sound we hear, the speed of a moving galaxy affects the colour of its light. This is called "red shift" and makes the most distant galaxies look much redder than they are. Despite this, the blobby galaxies of the Hubble Ultra-Deep Field are actually bluer than the galaxies of today's cosmos. Most of the earliest galaxies were irregular, and it was only as the universe grew older and settled down that spirals and ellipticals began to form. Quasars also become much more common at greater distances and further back in time, which means that they, too, are mostly echoes of the universe's violent early history.

Detail of 36 galaxies from the Hubble Ultra-Deep Field in various stages of formation.

THE EDGE OF EVERYTHING

THERE IS A LIMIT TO HOW FAR WE CAN SEE ACROSS THE UNIVERSE. OBJECTS MORE THAN 13.7 BILLION LIGHT YEARS AWAY ARE FOREVER HIDDEN FROM VIEW BECAUSE THEIR LIGHT HAS NOT HAD TIME TO REACH US SINCE THE BIG BANG. AROUND THE EDGE LIES A DARK ABYSS WHERE THE STARS HAVE NOT YET BEGUN TO SHINE. BEYOND THIS, ALL AROUND THE SKY, OUR RADIO TELESCOPES CAN DETECT A FAINT GLOW KNOWN AS THE COSMIC MICROWAVE BACKGROUND RADIATION (CMBR). THIS IS THE AFTERGLOW OF CREATION ITSELF, A REMNANT OF THE FIREBALL OF THE BIG BANG, COOLED TO JUST 3°C ABOVE ABSOLUTE ZERO BY ITS LONG JOURNEY TO REACH US.

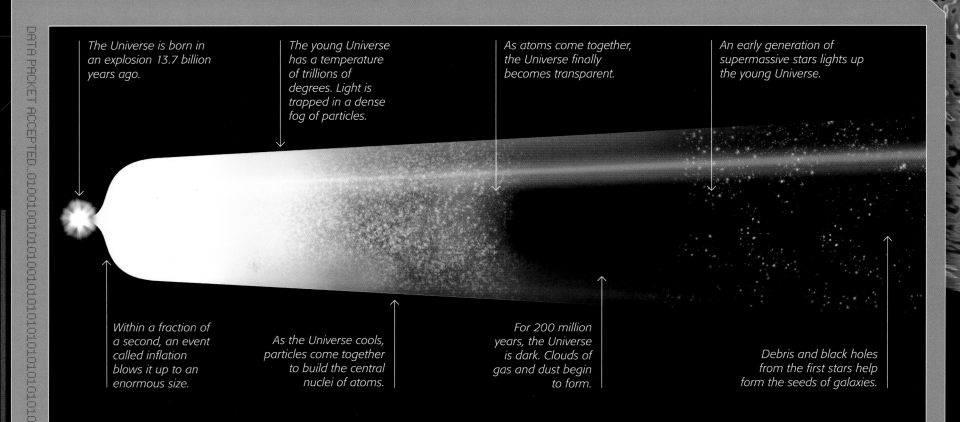

The Universe is born in an explosion 13.7 billion years ago.

The young Universe has a temperature of trillions of degrees. Light is trapped in a dense fog of particles.

As atoms come together, the Universe finally becomes transparent.

An early generation of supermassive stars lights up the young Universe.

Within a fraction of a second, an event called inflation blows it up to an enormous size.

As the Universe cools, particles come together to build the central nuclei of atoms.

For 200 million years, the Universe is dark. Clouds of gas and dust begin to form.

Debris and black holes from the first stars help form the seeds of galaxies.

About 13.7 billion years ago, the Universe exploded into being as an unimaginably dense and hot ball of matter – an event called the Big Bang. Space and time themselves were bound up in the infant Universe, so it's meaningless to ask where the Big Bang happened or what came before it. As the fundamental forces of nature separated from a single primeval "superforce", they drove a sudden period of rapid expansion called inflation. Since inflation, the Universe has cooled and expanded further, while its matter has come together – first forming atomic nuclei, then atoms and gradually stars and galaxies.

« EXPAND SEARCH
> THE BIG BANG

INDEX

ACKNOWLEDGMENTS

Quercus Publishing Plc
21 Bloomsbury Square
London WC1A 2NS

This edition published 2008 for Index Books Ltd
Copyright © Quercus Publishing Plc 2008

A CIP catalogue record for this book is available from the British Library

978-1-84724-775-9 (UK trade edition)
978-1-84724-524-3 (hardback edition)
978-1 84724-753-7 (paperback edition)

Printed in China

Created for Quercus by Tall Tree Ltd
Editors: Jon Richards and Kate Simkins
Designers: Ed Simkins, Jonathan Vipond and Ben Ruocco

Every effort has been made to contact copyright holders. However,
the publishers will be glad to rectify in future editions any inadvertent
omissions brought to their attention.

All images were supplied courtesy of NASA, except:
6l, 10–11, 12tl, 16tl, 20tl, 24tl, 26tl, 30tl, 38tl, 42tl, 44tl, 46tl, 47br, 48tl,
56tl, 60tl, 62tl, 64tl, 68–69, 81b, 86–87, 98–99, 104–105, 134–135,
140b, 141b Pikaia Imaging

Tall Tree Ltd would like to thank Chris Bernstein for the index.

Each new generation of stars is shorter-lived than the last.

If the Universe has enough mass, the expansion will eventually start to slow down.

If dark energy continues to increase its strength, the Universe might be torn apart in a "Big Rip".

A Universe that continues to expand steadily will die in a "Big Chill" as the stars go out and matter disintegrates.

Whatever its fate, the Universe will continue to expand for many billions of years.

However, thanks to dark energy, it's more likely that the Universe's expansion will accelerate.

With enough mass and a weak enough expansion, the Universe might eventually fall back in on itself.

A collapsing Universe could end in a "Big Crunch", which might in turn create a new Big Bang.

For a long time, astronomers thought the ultimate fate of the Universe depended solely on the amount of matter, both invisible and dark, that it contained. With enough mass in the Universe, gravity would eventually slow down the cosmic expansion and perhaps even pull the Universe back together in a "Big Crunch". With too little mass, it would keep expanding forever, growing colder as the stars faded and went out. Today, we know that another force called "dark energy" is acting against gravity and will probably keep the universe expanding forever, perhaps at an ever-increasing rate.

<< EXPAND SEARCH
> THE END?

GLOSSARY

Active galaxy
A galaxy that produces large amounts of energy as matter falls into a supermassive black hole at its centre. Depending on their appearance, active galaxies are also called quasars, blazars, radio galaxies and Seyfert galaxies.

Asteroid
A small rocky world of the inner solar system, usually found orbiting between Mars and Jupiter.

Atmosphere
A shell of gases held around a planet or star by its gravity.

Barred spiral galaxy
A spiral galaxy whose spiral arms are linked to its hub by a straight bar of stars and other material.

Binary star
A pair of stars that form from the same starbirth nebula and remain in orbit around one another throughout their lives.

Black hole
A superdense point in space formed by a collapsing stellar core more than five times the mass of the Sun. A black hole's gravity is so powerful that even light cannot escape from it.

Brown dwarf
A "failed star" – a ball of gas that never grew big enough to start shining by nuclear fusion, but which still glows dimly in infrared.

Comet
A chunk of rock and ice from the outer edge of the solar system. When a comet comes close to the Sun, it heats up and its surface ices evaporate, forming a gassy coma and a tail.

Core
The central region of a star or planet. In a star, the core is the region where temperatures and pressures are high enough to trigger nuclear fusion.

Dark nebula
A cloud of interstellar gas and dust that absorbs light, and is only visible when silhouetted against a cloud of stars or another type of nebula.

Dwarf planet
Any object that is in an independent orbit around the Sun, and has enough mass to pull itself into a roughly spherical shape, but which, unlike a true planet, has not cleared the region around it of other objects. Currently there are three known dwarf planets – the asteroid Ceres, and the Kuiper Belt objects Pluto and Eris.

Electromagnetic radiation
A form of energy combining electric and magnetic waves, taking many forms including light, emitted by many objects in the Universe.

Elliptical galaxy
A galaxy consisting of stars in random orbits, and lacking in star-forming gas. Ellipticals range in size between the smallest and largest galaxies known.

Flare
A huge release of superheated particles above the surface of a star, caused by a short-circuit in its magnetic field.

Galaxy
An isolated cluster of stars, gas and dust, often with a complex structure.

Gamma rays
The highest-energy form of electromagnetic radiation, generated by the hottest objects and most violent events in the Universe.

Giant planet
A planet with a huge envelope of gas, liquid or slushy ice (various frozen chemicals), around a relatively small solid core.

Globular cluster
A dense ball of ancient stars in orbit around a galaxy such as the Milky Way.

Infrared
Electromagnetic radiation with a little less energy than visible light. Infrared is often emitted by warm objects too cool to shine visibly.

Irregular galaxy
A galaxy with no obvious shape, usually rich in gas, dust and star-forming regions.

Kuiper Belt
A doughnut-shaped ring of icy worlds directly beyond the orbit of Neptune, filled with small icy worlds and comets. The largest known Kuiper Belt objects are Pluto and Eris.

Light year
The distance travelled by light (or other electromagnetic radiation) in one year. A light year is equivalent to roughly 9.5 million million kilometers.

Luminosity
A measure of the energy output of a star, usually compared to the Sun.

Main sequence
A term used to describe the longest phase in a star's life, when it is relatively stable and shines by fusing hydrogen into helium at its core. During this period, the star obeys rules linking its mass, size, luminosity and colour.

Multiple star
A system of two or more stars in orbit around one another (pairs of stars are also called binaries). Most of the stars in our galaxy are members of multiple systems rather than individuals like the Sun.

Nebula
Any cloud of gas or dust floating in space, ranging from the huge clouds in which stars are born, to spherical shells of planetary nebulae.

Neutron star
The collapsed core of a heavyweight star, left behind by a supernova explosion. A neutron star is the densest known object – though in the most massive stars, the core can collapse past the neutron star stage into a black hole. Many neutron stars are also pulsars.

Nova
A binary star system in which a white dwarf is pulling material away from a companion star, building up a layer of gas around itself that then burns away in a violent nuclear explosion.

Nuclear fusion
The process by which the stars shine. Fusion involves joining together light atomic nuclei (the central cores of atoms) to make heavier ones at very high temperatures and pressures, releasing energy in the process.

Oort Cloud
A spherical shell of dormant comets, perhaps two light years across, surrounding the solar system.

Open cluster
A large group of young stars that has recently been born in a star-forming nebula, and may still be embedded in its gas clouds.

Planet
An object that follows its own orbit around the Sun, has enough mass to pull itself into a spherical shape, and which has cleared the space around it of other objects (apart from satellites). According to this definition, there are eight planets – Mercury, Venus, Earth, Mars, Jupiter, Saturn, Uranus and Neptune.

Planetary nebula
An expanding cloud of glowing gas puffed off by a dying red giant star as it becomes a white dwarf.

Pulsar
A rapidly spinning neutron star with an intense magnetic field that channels its radiation out along two narrow beams that sweep across the sky.

Radio
The lowest-energy form of electromagnetic radiation. Radio waves are emitted by cool gas clouds in space, but also by violent active galaxies and pulsars.

Red dwarf
A star considerably lighter than the Sun – small, faint and with a low surface temperature. Red dwarfs live for much longer than the Sun, despite their tiny size.

Red giant
A star passing through a phase of its life where its luminosity has increased hugely, its outer layers have expanded and its surface has cooled. Stars enter a red giant phase as they exhaust the fuel supply in their core.

Reflection nebula
A cloud of interstellar gas and dust that shines by reflecting or scattering light from nearby stars.

Rocky planet
A relatively small planet composed largely of rocks and minerals, perhaps surrounded by a thin envelope of gas and liquid.

Solar wind
A stream of high-energy particles blown off the surface of the Sun by the pressure of its radiation, and spreading across the surrounding space.

Spectrum
The spread-out band of light created by passing light through a prism or other device. The prism bends light by different amounts depending on its colour, so the spectrum reveals the intensity of an object's light in different colours – a useful clue to the chemicals it contains.

Spiral galaxy
A galaxy consisting of a hub of old yellow stars, surrounded by a flattened disk of younger stars, gas and dust, with bright spiral arms marking regions of current star formation.

Star
A dense ball of gas that is hot and dense enough at its centre to trigger nuclear fusion reactions, causing it to shine.

Starbirth nebula
A glowing cloud of gas in space that gives birth to stars. Nebulae glow as their atoms are excited by high-energy light from nearby young stars.

Sun
The star at the centre of Earth's solar system. The Sun is a fairly average low-mass star, and a useful comparison for other stars.

Sunlike star
A yellow star with roughly the same mass, luminosity and surface temperature as the Sun.

Supergiant
A massive and extremely luminous star with 10 to 70 times the mass of the Sun. Supergiants range in colour from blue to red.

Supermassive black hole
A black hole with the mass of millions of stars, believed to lie in the very centre of many galaxies.

Supernova
A cataclysmic explosion marking the death of a star. Supernovae are triggered when a heavyweight star exhausts the last of its fuel and its core collapses into a neutron star or black hole. They can also form when a white dwarf in a nova system picks up enough mass to collapse suddenly into a neutron star.

Supernova remnant
A cloud of superheated gas expanding from the site of a former supernova explosion.

Ultraviolet
Electromagnetic radiation with wavelengths slightly shorter than visible light, usually radiated by objects hotter than the Sun.

Visible light
Electromagnetic radiation that can be detected with the human eye. Stars like the Sun emit most of their energy as visible light.

White dwarf
A stellar remnant left behind by the death of a star like the Sun. White dwarfs are the dense, slowly cooling cores of stars – still glowing with intense heat, but hard to see because of their tiny size.

X-rays
High-energy electromagnetic radiation emitted by extremely hot objects and violent processes in the Universe. Material heated as it is pulled towards a black hole is one of the strongest sources of astronomical x-rays.